Recent Titles in This Series

494 **Igor B. Frenkel, Yi-Zhi Huang, and James Lepowsky,** [illegible] approaches to vertex operator algebras and modules, 1993

493 **Nigel J. Kalton,** Lattice structures on Banach spaces, 1993

492 **Theodore G. Faticoni,** Categories of modules over endomorphism rings, 1993

491 **Tom Farrell and Lowell Jones,** Markov cell structures near a hyperbolic set, 1993

490 **Melvin Hochster and Craig Huneke,** Phantom homology, 1993

489 **Jean-Pierre Gabardo,** Extension of positive-definite distributions and maximum entropy, 1993

488 **Chris Jantzen,** Degenerate principal series for symplectic groups, 1993

487 **Sagun Chanillo and Benjamin Muckenhoupt,** Weak type estimates for Cesaro sums of Jacobi polynomial series, 1993

486 **Brian D. Boe and David H. Collingwood,** Enright-Shelton theory and Vogan's problem for generalized principal series, 1993

485 **Paul Feit,** Axiomization of passage from "local" structure to "global" object, 1993

484 **Takehiko Yamanouchi,** Duality for actions and coactions of measured groupoids on von Neumann algebras, 1993

483 **Patrick Fitzpatrick and Jacobo Pejsachowicz,** Orientation and the Leray-Schauder theory for fully nonlinear elliptic boundary value problems, 1993

482 **Robert Gordon,** G-categories, 1993

481 **Jorge Ize, Ivar Massabo, and Alfonso Vignoli,** Degree theory for equivariant maps, the general S^1-action, 1992

480 **L. Š. Grinblat,** On sets not belonging to algebras of subsets, 1992

479 **Percy Deift, Luen-Chau Li, and Carlos Tomei,** Loop groups, discrete versions of some classical integrable systems, and rank 2 extensions, 1992

478 **Henry C. Wente,** Constant mean curvature immersions of Enneper type, 1992

477 **George E. Andrews, Bruce C. Berndt, Lisa Jacobsen, and Robert L. Lamphere,** The continued fractions found in the unorganized portions of Ramanujan's notebooks, 1992

476 **Thomas C. Hales,** The subregular germ of orbital integrals, 1992

475 **Kazuaki Taira,** On the existence of Feller semigroups with boundary conditions, 1992

474 **Francisco González-Acuña and Wilbur C. Whitten,** Imbeddings of three-manifold groups, 1992

473 **Ian Anderson and Gerard Thompson,** The inverse problem of the calculus of variations for ordinary differential equations, 1992

472 **Stephen W. Semmes,** A generalization of riemann mappings and geometric structures on a space of domains in $\mathbf{C}^n$, 1992

471 **Michael L. Mihalik and Steven T. Tschantz,** Semistability of amalgamated products and HNN-extensions, 1992

470 **Daniel K. Nakano,** Projective modules over Lie algebras of Cartan type, 1992

469 **Dennis A. Hejhal,** Eigenvalues of the Laplacian for Hecke triangle groups, 1992

468 **Roger Kraft,** Intersections of thick Cantor sets, 1992

467 **Randolph James Schilling,** Neumann systems for the algebraic AKNS problem, 1992

466 **Shari A. Prevost,** Vertex algebras and integral bases for the enveloping algebras of affine Lie algebras, 1992

465 **Steven Zelditch,** Selberg trace formulae and equidistribution theorems for closed geodesics and Laplace eigenfunctions: finite area surfaces, 1992

464 **John Fay,** Kernel functions, analytic torsion, and moduli spaces, 1992

463 **Bruce Reznick,** Sums of even powers of real linear forms, 1992

(Continued in the back of this publication)

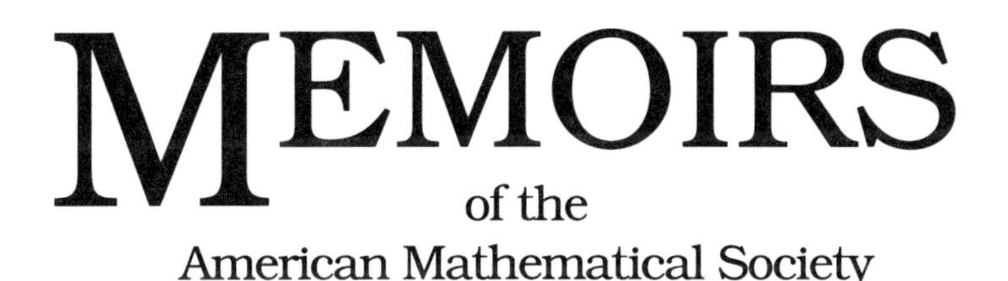

Number 494

On Axiomatic Approaches to Vertex Operator Algebras and Modules

Igor B. Frenkel
Yi-Zhi Huang
James Lepowsky

July 1993 • Volume 104 • Number 494 (first of 6 numbers) • ISSN 0065-9266

American Mathematical Society
Providence, Rhode Island

1991 *Mathematics Subject Classification.*
Primary 17xx, 17B68, 81T40.

Library of Congress Cataloging-in-Publication Data

Frenkel, Igor.
On axiomatic approaches to vertex operator algebras and modules/Igor B. Frenkel, Yi-Zhi Huang, James Lepowsky.
p. cm. – (Memoirs of the American Mathematical Society, ISSN 0065-9266; no. 494)
"Volume 104, number 494 (first of 6 numbers)."
Includes bibliographical references.
ISBN 0-8218-2555-0
1. Vertex operator algebras. 2. Modules (Algebra) I. Huang, Yi-Zhi, 1959– . II. Lepowsky, J. (James) III. Title. IV. Series.
QA3.A57 no. 494
[QA326]
510 s–dc20 93-17169
[512′.55] CIP

Memoirs of the American Mathematical Society

This journal is devoted entirely to research in pure and applied mathematics.

Subscription information. The 1993 subscription begins with Number 482 and consists of six mailings, each containing one or more numbers. Subscription prices for 1993 are $336 list, $269 institutional member. A late charge of 10% of the subscription price will be imposed on orders received from nonmembers after January 1 of the subscription year. Subscribers outside the United States and India must pay a postage surcharge of $25; subscribers in India must pay a postage surcharge of $43. Expedited delivery to destinations in North America $30; elsewhere $92. Each number may be ordered separately; *please specify number* when ordering an individual number. For prices and titles of recently released numbers, see the New Publications sections of the *Notices of the American Mathematical Society.*

Back number information. For back issues see the *AMS Catalog of Publications.*

Subscriptions and orders should be addressed to the American Mathematical Society, P. O. Box 1571, Annex Station, Providence, RI 02901-1571. *All orders must be accompanied by payment.* Other correspondence should be addressed to Box 6248, Providence, RI 02940-6248.

Memoirs of the American Mathematical Society is published bimonthly (each volume consisting usually of more than one number) by the American Mathematical Society at 201 Charles Street, Providence, RI 02904-2213. Second-class postage paid at Providence, Rhode Island. Postmaster: Send address changes to Memoirs, American Mathematical Society, P. O. Box 6248, Providence, RI 02940-6248.

Printed in the United States of America.
This volume was printed directly from author-prepared copy.
The paper used in this book is acid-free and falls within the guidelines established to ensure permanence and durability. ♾
10 9 8 7 6 5 4 3 2 1 98 97 96 95 94 93

CONTENTS

ABSTRACT

The basic definitions and properties of vertex operator algebras, modules, intertwining operators and related concepts are presented, following a fundamental analogy with Lie algebra theory. The first steps in the development of the general theory are taken, and various natural and useful reformulations of the axioms are given. In particular, it is shown that the Jacobi(-Cauchy) identity for vertex operator algebras – the main axiom – is equivalent (in the presence of more elementary axioms) to rationality, commutativity and associativity properties of vertex operators, and in addition, that commutativity implies associativity. These "duality" properties and related properties of modules are crucial in the axiomatic formulation of conformal field theory. Tensor product modules for tensor products of vertex operator algebras are considered, and it is proved that under appropriate hypotheses, every irreducible module for a tensor product algebra decomposes as the tensor product of (irreducible) modules. The notion of contragredient module is formulated, and it is shown that every module has a natural contragredient with certain basic properties. Adjoint intertwining operators are defined and studied. Finally, most of the ideas developed here are used to establish "duality" results involving two module elements, in a natural setting involving a module with integral weights.

Key words and phrases. Vertex operator algebras, Jacobi(-Cauchy) identity, Virasoro algebra, duality for vertex operator algebras, modules for vertex operator algebras, intertwining operators for vertex operator algebras, conformal field theory.

HISTORICAL NOTE

This paper was distributed as a preprint starting in 1989. Except for a few minor corrections, it is identical to the original preprint. The paper develops the basic axiomatic theory of vertex operator algebras; some of this material was already incorporated in the book "Vertex operator algebras and the Monster" [FLM] (1988), as was indicated there. Since then, these ideas have been applied in several directions by a number of people, and many new examples of vertex operator algebras and related structures have been studied. The importance of the category of vertex operator algebras has become more evident. We believe that the publication of this paper at the present time is no less useful than when it was written.

1. INTRODUCTION

Vertex operator algebras are a new and fundamental class of algebraic structures which have recently arisen in mathematics and physics. Their definition and some elementary properties and basic examples were presented in the book [FLM]. The goal of this paper is to lay some further foundations of the theory of vertex operator algebras and their representations.

The importance of these new algebras is supported by their numerous relations with other directions in mathematics and physics, such as the representation theory of the Virasoro algebra and affine Lie algebras, the theory of Riemann surfaces, knot invariants and invariants of three-dimensional manifolds, quantum groups, monodromy associated with differential equations, and conformal and topological field theories. In fact, the theory of vertex operator algebras can be thought of as an algebraic foundation of a great number of constructions in these theories.

The main original motivation for the introduction of the notion of vertex operator algebra arose from the problem of realizing the Monster sporadic group as a symmetry group of a certain infinite-dimensional graded vector space with natural additional structure. (See the Introduction in [FLM] for a historical discussion, including the important role of Borcherds' announcement [B].) The additional structure can be expressed in terms of the axioms defining these new algebraic objects (which are not actually algebras, even nonassociative algebras, in the usual sense). The Monster is in fact the symmetry group of a special vertex operator algebra, the moonshine module, just as the Mathieu group M_{24} is the symmetry group of a special error-correcting code, the Golay code, and the Conway group Co_0 is the symmetry group of a special positive definite even lattice, the Leech lattice. All three special objects possess and can be characterized by the following properties (the uniqueness being conjectural in the Monster case):

(a) "self-dual"
(b) "rank 24"
(c) "no small elements,"

which have appropriate definitions for each of the three types of mathematical structures. In the case of vertex operator algebras the notion of self-duality means that there is only one irreducible module (the moonshine module itself). Thus even apart from other concepts in mathematics and physics, the Monster alone leads to the notions of vertex operator algebras and their representations.

Received by editor September 2, 1991.

In the physics literature the main ingredients of the definition of the physical counterpart of vertex operator algebras were discovered in relation first to the dual resonance model and then to conformal field theory (see for instance the Introduction in [FLM] for a discussion of the history). One of the focal points of the axiomatic formulation of conformal field theory was the paper [BPZ], in which the role of the Virasoro algebra was especially emphasized. The modern notion of chiral algebra accepted now in the physics literature essentially coincides with our notion of vertex operator algebra; see e.g. [MS]. In particular, the mutual locality, or "commutativity," of operators and the "associativity" of the operator product expansion are necessary properties of chiral algebras. Our argument that the latter follows from the former under certain natural conditions can be used to simplify the verification of the axioms in concrete examples [FLM]; see also [G]. Many important discoveries involving representations of chiral algebras and the associated intertwining operators, which are known in the physics literature as chiral vertex operators, have recently been made in such works as [TK], [V] and [MS]. The latter paper extends and develops the axiomatic approach to conformal field theory, and it also contains an extensive review of the relevant physics literature in the five-year period since [BPZ].

The present paper starts from a rigorous definition of vertex operator algebra (the same as that introduced in [FLM]), a definition implicit, but not completely explicit, in the physics literature, and it serves the purpose of building a foundation for the rich structures associated to conformal field theory and mentioned above. The scope of this work is to present the "monodromy-free" fundamentals and basic results of a rapidly-developing theory; we treat the situations in which the matrix coefficients of compositions of vertex operators are essentially single-valued rational functions. This paper overlaps, elaborates and extends the axiomatic material presented in [FLM], especially in Chapter 8 and the Appendix.

Here we explain some of the basic axioms of vertex operator algebras and their relation to classical mathematical notions, in particular, to Lie algebras. Let V be a vector space over a field $\mathbb{F}$, assumed for later purposes to have characteristic 0, and let

$$\mathrm{ad}(\cdot)\,\cdot : V \otimes V \to V \tag{1.1}$$

be a linear map satisfying the identity

$$\mathrm{ad}(u)\mathrm{ad}(v) - \mathrm{ad}(v)\mathrm{ad}(u) = \mathrm{ad}(\mathrm{ad}(u)v) \tag{1.2}$$

for any $u, v \in V$. Then if we require that

$$\mathrm{ad}(v) = 0 \quad \text{implies} \quad v = 0, \tag{1.3}$$

the pair (V, ad) is nothing but a Lie algebra having zero center, with ad denoting the adjoint representation, i.e., the Lie bracket is given by:

$$[u, v] = \mathrm{ad}(u)v. \tag{1.4}$$

Then (1.2) is one of the equivalent forms of the Jacobi identity, which together with (1.3) also implies the skew-symmetry of the bracket.

The above form of the Jacobi identity is also parallel to the definition of Lie algebra representation. In fact, by a representation of the Lie algebra V on the module W one understands a linear map

$$\pi(\cdot)\cdot : V \otimes W \to W \tag{1.5}$$

satisfying the identity

$$\pi(u)\pi(v) - \pi(v)\pi(u) = \pi(\mathrm{ad}(u)v) \tag{1.6}$$

for any $u, v \in V$. To round out the basic notions of Lie algebra and representations one defines the tensor product of two modules (W_1, π_1), (W_2, π_2) and then the notion of intertwining operator from their tensor product to a third module (W_3, π_3) :

$$I(\cdot)\cdot : W_1 \otimes W_2 \to W_3, \tag{1.7}$$

which we also put into a form similar to (1.2) and (1.6):

$$\pi_3(u)I(v) - I(v)\pi_2(u) = I(\pi_1(u)v). \tag{1.8}$$

Note that the module structure on the space $W_1 \otimes W_2$ does not enter into this formula. Starting from these definitions one can proceed to study representation theory.

The theory of vertex operator algebras can be developed in a completely parallel way. Each of the three definitions and identities has its vertex operator algebra analogue which contains as an additional ingredient the Cauchy residue formula, which may be written as:

$$-\mathrm{Res}_{z=\infty} f(z) - \mathrm{Res}_{z=0} f(z) = \mathrm{Res}_{z=z_0} f(z). \tag{1.9}$$

Here we take $f(z)$ to be a rational function of one complex variable with its only poles at 0, ∞ and z_0, and we observe that this formula makes perfect sense over our field $\mathbb{F}$. Let V be a vector space over $\mathbb{F}$ and let

$$\mathrm{ad}_z(\cdot)\cdot : V \otimes V \to V((z)) \tag{1.10}$$

be a linear map, where $V((z))$ denotes the algebra of those formal Laurent series in the formal variable z involving at most finitely many negative powers of z. Then ad_z is the generating function of an infinite family of linear maps from $V \otimes V$ to V. The main axiom for a vertex operator algebra is what we call the *Jacobi-Cauchy identity*, or, especially in the alternative version given in the main text of this paper, the *Jacobi identity* for vertex operator algebras:

$$\begin{aligned} -\mathrm{Res}_{z=\infty}\left(f(z)\mathrm{ad}_z(u)\mathrm{ad}_{z_0}(v)\right) - \mathrm{Res}_{z=0}\left(f(z)\mathrm{ad}_{z_0}(v)\mathrm{ad}_z(u)\right) \\ = \mathrm{Res}_{z=z_0}\left(f(z)\mathrm{ad}_{z_0}(\mathrm{ad}_{z-z_0}(u)v)\right), \end{aligned} \tag{1.11}$$

where f is as above and where the residues are defined, as for scalar-valued rational functions or formal series, as certain coefficients in appropriate expansions.

The precise meaning of formula (1.11), including the roles of operator-valued rational functions and of their various formal series expansions, will be discussed in the main text. In particular, the single identity (1.11) is equivalent to an infinite family of identities for the component operators (see Corollary 8.8.17 in [FLM]).

The definitions of a representation π_z of the vertex operator algebra V and of intertwining operators I_z between representations are again formed by combining the corresponding definitions for Lie algebras with the Cauchy residue formula (1.9). In the main text, we shall use the notations $Y(v,z)$ for both $\mathrm{ad}_z(v)$ and $\pi_z(v)$, and the notation $\mathcal{Y}(v,z)$ for $I_z(v)$. These operators are all called vertex operators, in their appropriate settings. (The algebra and module operators $Y(\cdot,z)$ are fixed, while the intertwining operators $\mathcal{Y}(\cdot,z)$ for a given triple of modules form a vector space.) The Jacobi identity for algebras, modules or intertwining operators is a precise statement of what is known in the physics literature as the Ward identity on the three-punctured Riemann sphere, and our choice of the notation Y for vertex operators reflects the shape of the tree diagram associated with this geometric picture.

One of the basic features of the Jacobi identity for Lie algebras, modules and intertwining operators is its symmetry with respect to the symmetric group $\mathcal{S}_3$. To show an analogous symmetry in the case of vertex operator algebras, modules and intertwining operators, it is useful to display first an $\mathcal{S}_3$-symmetry of the Cauchy residue formula. For any algebraic expression z, we set $\delta(z)=\sum_{n\in\mathbb{Z}}z^n$ provided that this sum makes sense. This expression is the formal analogue of the δ-distribution at $z=1$; in particular, $\delta(z)f(z)=\delta(z)f(1)$ for any f for which these expressions are defined. Consider the following formal Laurent series in three commuting variables:

$$z_0^{-1}\delta\left(\frac{z_1-z_2}{z_0}\right)=z_0^{-1}\sum_{n\in\mathbb{Z}}\frac{(z_1-z_2)^n}{z_0^n},\tag{1.12}$$

where we expand the negative powers $(z_1-z_2)^n$ as formal power series in the second variable, z_2. Then the residue formula (1.9) can be expressed in terms of such formal series as follows:

$$\begin{aligned}z_0^{-1}\delta\left(\frac{z_1-z_2}{z_0}\right)F(z_0,z_1,z_2)+z_2^{-1}\delta\left(\frac{z_0-z_1}{-z_2}\right)F(z_0,z_1,z_2)\\=z_1^{-1}\delta\left(\frac{z_2+z_0}{z_1}\right)F(z_0,z_1,z_2),\end{aligned}\tag{1.13}$$

where F is a Laurent polynomial in z_0,z_1,z_2; in the first term, F can be replaced by $F(z_1-z_2,z_1,z_2)$, expanded as a Laurent series in z_2 involving only finitely many negative powers of z_2, and analogously for the two other terms, where we expand in large powers of z_1 and z_0, respectively. Thanks to the new variable z_0, the $\mathcal{S}_3$-symmetry properties become clear. (The symmetry becomes completely apparent if we replace z_1 by $-z_1$ and move the right-hand side to the left; the δ-functions then correspond to the symbolic relation $z_0+z_1+z_2=0$.)

In the Jacobi identity for vertex operator algebras, modules and intertwining operators, special expressions are used in the role of F; these expressions appeared symbolically in (1.11) and are discussed further below. This Jacobi identity has a natural $\mathcal{S}_3$-symmetry which might be thought of as a blending of the separate symmetries of the two classical formulas. The restoration of this symmetry by means of the introduction of the third variable and the elimination of the test function $f(z)$ in the Jacobi-Cauchy identity (1.11) are advantages of the formal-variable language that we use throughout the main text.

Besides the main axiom it is appropriate to introduce a few additional axioms which further specify the notion of vertex operator algebra. These axioms are natural from several points of view that we shall not discuss in this Introduction. (See instead the Introduction in [FLM].) By analogy with (1.3), we assume that

$$Y(v,z)=0 \quad \text{implies} \quad v=0 \quad \text{for} \quad v\in V. \tag{1.14}$$

We also require the existence of two special elements $\mathbf{1}$ and ω of V such that

$$Y(\mathbf{1},z)=1 \qquad \text{(the identity operator)} \tag{1.15}$$

$$Y(\omega,z)=\sum_{n\in\mathbb{Z}}L(n)z^{-n-2}, \tag{1.16}$$

where the $L(n)$ are the generators of the Virasoro algebra with a normalized central element acting as multiplication by a scalar called the *rank* of the vertex operator algebra. We also assume that

$$\frac{d}{dz}Y(v,z)=Y(L(-1)v,z) \tag{1.17}$$

and that V is $\mathbb{Z}$-graded, with the grading truncated from below and with the homogeneous subspaces finite-dimensional and equal to the corresponding eigenspaces of $L(0)$. In the presence of the other axioms, the injectivity (1.14) is equivalent to the *creation property*, which states that $Y(v,z)\mathbf{1}$ is a power series in z, and that its constant term is v. Appropriate related axioms define modules (representation spaces) and intertwining operators. One important difference between the vertex operator algebra case and the Lie algebra case (recall (1.7) and (1.8)) is that in the present case, the tensor product of modules for a vertex operator algebra does not in fact carry a module structure, but intertwining operators are well defined nevertheless. This completes our sketch of the definition of the concepts of vertex operator algebra and the related structures.

The three operators $L(-1)$, $L(0)$ and $L(1)$ (the latter playing the role of $L(-1)$ in the contragredient module; see below) are particularly important in the structure of vertex operator algebras. They span a subalgebra isomorphic to $\mathfrak{sl}(2)$, and in a suitable sense, the corresponding group acts by projective transformations on the variable z. One can weaken the vertex operator algebra axioms, postulating instead of the full Virasoro algebra only this three-dimensional subalgebra, with appropriate properties. We call such objects *quasi-vertex operator algebras.* There are a few other natural ways to modify the axiom system, as the results in this paper, and in [B] and [FLM], make clear. However, the present state of development of the theory naturally leads to the axiom system chosen here (and in [FLM]).

It is by no means obvious that nontrivial examples of vertex operator algebras exist. In fact, the construction of such objects is quite a long story, not treated in this paper, which is devoted instead to axiomatic considerations. For instance, the construction of the moonshine module for the Monster, together with related vertex operator algebras and modules, is presented in detail in [FLM].

As we have been discussing, vertex operator algebras are similar in spirit to Lie algebras. However, they are also similar in spirit to commutative associative algebras: The Jacobi identity is equivalent to two forms of "duality" (in the language of conformal field theory):

$$Y(u,z)Y(v,z_0) \sim Y(v,z_0)Y(u,z) \tag{1.18}$$

$$Y(u,z)Y(v,z_0) \sim Y(Y(u,z-z_0)v,z_0), \tag{1.19}$$

where the symbol $\sim$ is understood as an equality of the (operator-valued) rational functions corresponding to the formal Laurent series defined by the indicated expressions. In fact, the function F in (1.13) is taken to be an arbitrary matrix coefficient of this common rational function (cf. (1.11)). Moreover, the second form of duality (1.19) follows from the first (1.18) and properties of $L(-1)$ and $L(0)$, and this fact provides a useful approach to constructing vertex operator algebras. The second form of duality is known as the associativity of the operator product expansion in conformal field theory. Although vertex operator algebras are not actually associative algebras, such results as the existence of tensor products of vertex operator algebras and the factorizability of irreducible modules for such tensor products still hold in the vertex operator algebra context. On the other hand, an additional feature of the theory (besides the Jacobi identity) analogous to Lie algebra theory as opposed to associative algebra theory is the concept of contragredient module and the associated adjoint operators. These concepts involve the nontrivial use of $\mathfrak{sl}(2)$. It is also interesting that it is sometimes valuable, within a number of arguments, to pass back and forth between the equivalent axiom systems. For instance, to prove the Jacobi identity for a tensor product of vertex operator algebras, it is easiest to verify the duality relations instead, and to prove the appropriate general duality relations for modules, it is easiest to use the $\mathcal{S}_3$-symmetry of the Jacobi identity. As another example, the Jacobi identity is useful in extending commutativity to products of several vertex operators. The feature of vertex operator algebras that on the one hand they are very similar to the classical structures of Lie and associative algebras, and that on the other hand they possess enormously rich structure embracing many previously unrelated classical objects, as referred to above, is one of the most beautiful aspects of this new theory.

This paper is devoted to the basic properties of vertex operator algebras, their representations and their intertwining operators that we have mentioned. It is written in an essentially self-contained way, although a brief look at Chapters 2 and 8 of [FLM] might be helpful to the reader. We work in an algebraic setting over the general field $\mathbb{F}$ of characteristic 0. In Section 2 we give the definition of vertex operator algebra and some consequences, including the S_3-symmetry of the Jacobi identity. We introduce the standard kinds of elementary categorical notions appropriate to vertex operator algebras, including tensor products. We discuss conformal and projective transformations of vertex operators and we define and comment on the more general class of quasi-vertex operator algebras. In Section 3 we formulate and prove the basic duality properties of vertex operator algebras, namely, the rationality of the two kinds of compositions of vertex operators – products and iterations – and commutativity and associativity. Then we show that these rationality, commutativity and associativity properties may be used in place of the Jacobi identity in the definition of vertex operator algebras. Our treatment over the field $\mathbb{F}$ is based on a formalism of formal Laurent series expansions of rational functions. We show that the associativity follows from commutativity for products of two vertex operators together with certain other properties of vertex operator algebras. We extend rationality and commutativity to several variables, using the Jacobi identity; associativity could also be extended analogously to several variables. Finally, we apply the duality formulation of vertex operator algebras to prove that the tensor product of vertex operator algebras is again a vertex operator algebra (a result stated in Section 2).

In Section 4 we define modules for vertex operator algebras and present consequences of the definition, elementary categorical notions and duality properties, in parallel with the corresponding parts of Section 2. We introduce tensor product modules for tensor product algebras and prove that their irreducibility is equivalent to the irreducibility of the factor modules. We also establish that under a natural hypothesis, every irreducible module for a tensor product of algebras is in fact a tensor product of irreducible modules for the factors. In Section 5 we study the duality properties of modules in greater depth. First we motivate and construct the vertex operators which correspond to elements of the module and map from the vertex operator algebra to the module and then we prove the various forms of duality for one module element and two algebra elements. We also establish several converse statements which allow us to replace the Jacobi identity in the definition of module by any of various duality assumptions. We then define the notion of adjoint vertex operator (using a formula in [B]), acting on the suitably restricted dual space of a given module, and using this, together with appropriate $\mathfrak{sl}(2)$ relations, we prove that every module has a natural contragredient module. We show that these contragredient modules behave similarly to those in classical algebraic theories; in particular, the double contragredient module is canonically isomorphic to the original module, and irreducibility is preserved under contragredience. As in classical algebraic theories, we interpret nondegenerate "invariant" bilinear forms and pairings in terms of contragredient modules. Then we show that if a vertex operator algebra is self-contragredient as a module for itself, then the corresponding form is symmetric. Next we define intertwining operators and prove their basic properties. (Our definition is slightly different from that in [MS].) We introduce adjoint intertwining operators, prove that they are indeed intertwining operators, and give their basic properties. One important corollary is an S_3-symmetry involving the dimensions of spaces of intertwining operators. These dimensions are termed *fusion rules*, as in [V], because of their analogy with the multiplicities of irreducible modules in tensor products of irreducible modules. We end the section with the formulation and proof of the various forms of duality for two module elements and one algebra element. This is based on the use of adjoint intertwining operators to construct vertex operators which correspond to elements of a module and map from the module to the algebra. The proof of these duality relations uses most of the concepts and arguments presented throughout the paper. This concludes our treatment of the monodromy-free axiomatic foundation of the general theory of vertex operator algebras, modules and intertwining operators.

Much of this work, which is an outgrowth of the book [FLM], was done during the final stages of the writing of [FLM]. We would like to thank Robert Wilson for helpful comments. We gratefully acknowledge partial support from the following sources: I.F.: the Institute for Advanced Study, Institut des Hautes Etudes Scientifiques and National Science Foundation grants DMS-8602091 and DMS-8906772; Y.Z.H.: a Rutgers University Graduate Excellence Fellowship and a Sloan Foundation Doctoral Dissertation Fellowship; J.L.: a Guggenheim Foundation Fellowship, the Rutgers University Faculty Academic Study Program and National Science Foundation grant DMS-8603151.

Additional note: After this work was completed we received the preprint of the paper [DGM], which includes a new proof and generalization of one of the main theorems in [FLM]; some of the methods in [DGM] are similar to some of the material in the present paper.

2. VERTEX OPERATOR ALGEBRAS

2.1. Formal calculus

We shall work in an algebraic setting over an arbitrary field $\mathbb{F}$ of characteristic 0. The set of nonnegative integers will be denoted by $\mathbb{N}$. The symbols $z, z_0, z_1, ...$ will designate commuting formal variables. For a vector space V, we set

$$V[z] = \left\{ \sum_{n \in \mathbb{N}} v_n z^n | v_n \in V, \text{ all but finitely many } v_n = 0 \right\} \tag{2.1.1}$$

$$V[z, z^{-1}] = \left\{ \sum_{n \in \mathbb{Z}} v_n z^n | v_n \in V, \text{ all but finitely many } v_n = 0 \right\} \tag{2.1.2}$$

$$V[[z]] = \left\{ \sum_{n \in \mathbb{N}} v_n z^n | v_n \in V \right\} \tag{2.1.3}$$

$$V((z)) = \left\{ \sum_{n \in \mathbb{Z}} v_n z^n | v_n \in V, \ v_n = 0 \text{ for sufficiently small } n \right\}, \tag{2.1.4}$$

$$V[[z, z^{-1}]] = \left\{ \sum_{n \in \mathbb{Z}} v_n z^n | v_n \in V \right\}, \tag{2.1.5}$$

and we shall also use analogous notation for several variables.

The following formal version of Taylor's theorem is easily verified by direct expansion: For $f(z) = \sum v_n z^n \in V[[z, z^{-1}]]$,

$$e^{z_0 \frac{d}{dz}} f(z) = f(z + z_0), \tag{2.1.6}$$

where an expression of the form e^x denotes the formal exponential series, and where on the right-hand side, each binomial $(z + z_0)^n$ is understood to be expanded in nonnegative integral powers of the second summand, namely, z_0. We shall repeatedly use this binomial expansion convention.

We introduce a basic generating function, the "formal δ-function at $z = 1$" :

$$\delta(z) = \sum_{n \in \mathbb{Z}} z^n. \tag{2.1.7}$$

The fundamental property of the δ-function is:

$$f(z)\delta(z) = f(1)\delta(z) \text{ for } f(z) \in \mathbb{F}[z, z^{-1}], \tag{2.1.8}$$

proved trivially by observing its truth for $f(z) = z^n$ and using linearity. This property has many variants; in general, whenever an expression is multiplied by the δ-function, we may formally set the argument appearing in the δ-function equal to 1, provided the relevant algebraic expressions make sense. For instance, given a formal Laurent series in two variables

$$X(z_1, z_2) \in (\text{End } W)[[z_1, z_1^{-1}, z_2, z_2^{-1}]] \tag{2.1.9}$$

with coefficients which are operators on a vector space W, such that

$$\lim_{z_1 \to z_2} X(z_1, z_2) \text{ exists, i.e., } X(z_1, z_2)|_{z_1 = z_2} \text{ exists} \tag{2.1.10}$$

(that is, when $X(z_1, z_2)$ is applied to any element of W, setting the variables equal leads to only finite sums in W), we have

$$X(z_1, z_2)\delta\left(\frac{z_1}{z_2}\right) = X(z_2, z_2)\delta\left(\frac{z_1}{z_2}\right). \tag{2.1.11}$$

As in this case, all limits, products of formal Laurent series and other operations will be understood in a purely algebraic sense.

In the theory of vertex operator algebras we shall often use three-variable generating functions of the following sort:

$$z_0^{-1}\delta\left(\frac{z_1 - z_2}{z_0}\right) = \sum_{n \in \mathbb{Z}} \frac{(z_1 - z_2)^n}{z_0^{n+1}} = \sum_{m \in \mathbb{N},\, n \in \mathbb{Z}} (-1)^m \binom{n}{m} z_0^{-n-1} z_1^{n-m} z_2^m. \tag{2.1.12}$$

There are two basic properties of the δ-function involving such expressions:

$$z_1^{-1}\delta\left(\frac{z_2 + z_0}{z_1}\right) = z_2^{-1}\delta\left(\frac{z_1 - z_0}{z_2}\right) \tag{2.1.13}$$

$$z_0^{-1}\delta\left(\frac{z_1 - z_2}{z_0}\right) - z_0^{-1}\delta\left(\frac{z_2 - z_1}{-z_0}\right) = z_2^{-1}\delta\left(\frac{z_1 - z_0}{z_2}\right). \tag{2.1.14}$$

These are easily proved by direct expansion. Note that the three terms in (2.1.14) are formal power series in (i.e., involve only nonnegative integral powers of) z_2, z_1 and z_0, respectively. In particular, the two terms on the left-hand side are unequal formal Laurent series in three variables, even though at first glance they appear equal.

The following residue notation will be useful:

$$\text{Res}_z \left(\sum_{n \in \mathbb{Z}} v_n z^n\right) = v_{-1}. \tag{2.1.15}$$

For instance, for the expression in (2.1.13),

$$\text{Res}_{z_1} z_1^{-1}\delta\left(\frac{z_2 + z_0}{z_1}\right) = 1. \tag{2.1.16}$$

2.2. Definition of vertex operator algebras

DEFINITION 2.2.1. A *vertex operator algebra* is a $\mathbb{Z}$-graded vector space (graded by *weights*)

$$V = \coprod_{n\in\mathbb{Z}} V_{(n)};\quad \text{for}\ \ v \in V_{(n)},\ \ n = \text{wt}\ v; \tag{2.2.1}$$

such that

$$\dim\ V_{(n)} < \infty \ \text{for}\ n \in \mathbb{Z}, \tag{2.2.2}$$

$$V_{(n)} = 0 \ \ \text{for}\ n\ \text{sufficiently small}, \tag{2.2.3}$$

equipped with a linear map $V \otimes V \to V[[z, z^{-1}]]$, or equivalently,

$$\begin{aligned} V &\to (\text{End}\ V)[[z, z^{-1}]] \\ v &\mapsto Y(v, z) = \sum_{n\in\mathbb{Z}} v_n z^{-n-1}\ (\text{where}\ v_n \in \text{End}\ V), \end{aligned} \tag{2.2.4}$$

$Y(v, z)$ denoting the *vertex operator associated with* v, and equipped also with two distinguished homogeneous vectors $\mathbf{1}$ (the *vacuum*) and $\omega \in V$. The following conditions are assumed for $u, v \in V$:

$$u_n v = 0 \ \ \text{for}\ n\ \text{sufficiently large}; \tag{2.2.5}$$

$$Y(\mathbf{1}, z) = 1 \ \ (1\ \text{on the right being the identity operator}); \tag{2.2.6}$$

the *creation property* holds:

$$Y(v, z)\mathbf{1} \in V[[z]] \quad \text{and} \quad \lim_{z\to 0} Y(v, z)\mathbf{1} = v \tag{2.2.7}$$

(that is, $Y(v, z)\mathbf{1}$ involves only nonnegative integral powers of z and the constant term is v); the *Jacobi identity* holds:

$$\begin{aligned} z_0^{-1}\delta\left(\frac{z_1 - z_2}{z_0}\right)Y(u, z_1)Y(v, z_2) - z_0^{-1}\delta\left(\frac{z_2 - z_1}{-z_0}\right)Y(v, z_2)Y(u, z_1) \\ = z_2^{-1}\delta\left(\frac{z_1 - z_0}{z_2}\right)Y(Y(u, z_0)v, z_2) \end{aligned} \tag{2.2.8}$$

(note that when each expression in (2.2.8) is applied to any element of V, the coefficient of each monomial in the formal variables is a finite sum; on the right-hand side, the notation $Y(\cdot, z_2)$ is understood to be extended in the obvious way to $V[[z_0, z_0^{-1}]]$); the Virasoro algebra relations hold:

$$[L(m), L(n)] = (m - n)L(m + n) + \frac{1}{12}(m^3 - m)\delta_{m+n,0}(\text{rank}\ V) \ \ \text{for}\ \ m, n \in \mathbb{Z}, \tag{2.2.9}$$

where

$$L(n) = \omega_{n+1} \text{ for } n \in \mathbb{Z}, \text{ i.e., } Y(\omega, z) = \sum_{n\in\mathbb{Z}} L(n) z^{-n-2} \tag{2.2.10}$$

and

$$\text{rank } V \in \mathbb{F}; \tag{2.2.11}$$

$$L(0)v = nv = (\text{wt } v)v \text{ for } n \in \mathbb{Z} \text{ and } v \in V_{(n)}; \tag{2.2.12}$$

$$\frac{d}{dz} Y(v, z) = Y(L(-1)v, z). \tag{2.2.13}$$

This completes the definition. We may denote the vertex operator algebra just defined by

$$(V, Y, \mathbf{1}, \omega). \tag{2.2.14}$$

In practice we will typically have

$$\text{rank } V \in \mathbb{Q}, \quad \text{rank } V \geq 0. \tag{2.2.15}$$

Remark 2.2.2. Axioms (2.2.4) and (2.2.5) are together equivalent to a single axiom – that V be equipped with a linear map

$$\begin{aligned} V \otimes V &\to V((z)) \\ v_1 \otimes v_2 &\mapsto Y(v_1, z) v_2. \end{aligned} \tag{2.2.16}$$

Remark 2.2.3. Property (2.1.14) of the δ-function amounts to the case $u = v = \mathbf{1}$ of the Jacobi identity.

Remark 2.2.4. In the presence of the other axioms, we can replace the creation-property axiom (2.2.7), whose naturality will become apparent, by the natural injectivity condition

$$Y(v, z) = 0 \text{ implies } v = 0 \text{ for } v \in V. \tag{2.2.17}$$

To see that (2.2.7) and (2.2.17) are equivalent, first note that (2.2.17) follows immediately from (2.2.7). The following converse argument provides an excellent illustration of the methods of formal calculus. Assume that all the axioms except (2.2.7) hold, together with (2.2.17). Using the basic δ-function properties, the

Jacobi identity (2.2.8) with $v = \mathbf{1}$ and property (2.2.6), we get

$$
\begin{aligned}
&z_1^{-1}\delta\left(\frac{z_2+z_0}{z_1}\right)Y(u,z_2+z_0)\\
&= z_1^{-1}\delta\left(\frac{z_2+z_0}{z_1}\right)Y(u,z_1)\\
&= z_2^{-1}\delta\left(\frac{z_1-z_0}{z_2}\right)Y(u,z_1)\\
&= \left(z_0^{-1}\delta\left(\frac{z_1-z_2}{z_0}\right)-z_0^{-1}\delta\left(\frac{z_2-z_1}{-z_0}\right)\right)Y(u,z_1)\\
&= z_0^{-1}\delta\left(\frac{z_1-z_2}{z_0}\right)Y(u,z_1)-z_0^{-1}\delta\left(\frac{z_2-z_1}{-z_0}\right)Y(u,z_1)\\
&= z_2^{-1}\delta\left(\frac{z_1-z_0}{z_2}\right)Y(Y(u,z_0)\mathbf{1},z_2)\\
&= z_1^{-1}\delta\left(\frac{z_2+z_0}{z_1}\right)Y(Y(u,z_0)\mathbf{1},z_2).
\end{aligned}
$$

Taking Res_{z_1} (cf. (2.1.16)) we see that

$$Y(u,z_2+z_0) = Y(Y(u,z_0)\mathbf{1},z_2). \tag{2.2.18}$$

But Taylor's theorem (2.1.6) and the $L(-1)$-derivative axiom (2.2.13) give

$$Y(e^{z_0L(-1)}u,z_2) = e^{z_0\frac{d}{dz_2}}Y(u,z_2) = Y(u,z_2+z_0), \tag{2.2.19}$$

so that by the injectivity assumption (2.2.17),

$$Y(u,z_0)\mathbf{1} = e^{z_0L(-1)}u. \tag{2.2.20}$$

Now (2.2.7) follows immediately.

Remark 2.2.5. The Jacobi identity may of course be written in terms of the components v_n of the operators $Y(v,z)$; see e.g. [FLM].

2.3. Consequences of the definition

Some immediate consequences of the definition follow:

$$[L(-1),Y(v,z)] = Y(L(-1)v,z) \tag{2.3.1}$$

$$[L(0),Y(v,z)] = Y(L(0)v,z) + zY(L(-1)v,z) \tag{2.3.2}$$

$$[L(1),Y(v,z)] = Y(L(1)v,z) + 2zY(L(0)v,z) + z^2Y(L(-1)v,z) \tag{2.3.3}$$

$$L(n)\mathbf{1} = 0 \quad \text{for} \quad n \geq -1 \tag{2.3.4}$$

$$L(-2)\mathbf{1} = \omega \tag{2.3.5}$$

$$L(0)\omega = 2\omega. \tag{2.3.6}$$

Formula (2.3.4) implies that the vacuum vector $\mathbf{1}$ is annihilated by the operators $L(-1)$, $L(0)$, $L(1)$, which span a copy of $\mathfrak{sl}(2) = \mathfrak{sl}(2, \mathbb{F})$ (assuming that these operators are nonzero). We have

$$\text{wt } \mathbf{1} = 0 \tag{2.3.7}$$

$$\text{wt } \omega = 2. \tag{2.3.8}$$

From (2.2.4), (2.2.12), (2.2.13) and (2.3.2), we find that if $v \in V$ is homogeneous, then

$$\text{wt } v_n = \text{wt } v - n - 1 \tag{2.3.9}$$

as an operator. In particular, the operator $x_v(n)$ defined by

$$Y(v, z) = \sum_{n \in \mathbb{Z}} x_v(n) z^{-n-\text{wt } v} \tag{2.3.10}$$

when v is a homogeneous vector satisfies the condition

$$\text{wt } x_v(n) = -n. \tag{2.3.11}$$

The notation $x_v(n)$ may be extended from homogeneous v to arbitrary $v \in V$ by linearity, and (2.3.11) holds in general.

We know that $Y(v, z)$ determines v (2.2.17). The vacuum vector $\mathbf{1}$ is uniquely determined by (2.2.6) and this injectivity (or the creation property (2.2.7)).

Using the other properties, we see that the Virasoro algebra commutation relations (2.2.9) are equivalent to:

$$Y(\omega, z)\omega = \frac{1}{2}(\text{rank } V)\mathbf{1} z^{-4} + 2\omega z^{-2} + L(-1)\omega z^{-1} + v \tag{2.3.12}$$

where $v \in V[[z]]$.

Taking Res_{z_0} of the Jacobi identity and using (2.1.14), we obtain the commutator formula

$$\begin{aligned}[Y(u, z_1), Y(v, z_2)] &= \text{Res}_{z_0} z_2^{-1} \delta\left(\frac{z_1 - z_0}{z_2}\right) Y(Y(u, z_0)v, z_2) \\ &= Y((Y(u, z_1 - z_2) - Y(u, -z_2 + z_1))v, z_2).\end{aligned} \tag{2.3.13}$$

Observe that only the singular terms in the expression $Y(u, z_0)v$ (the terms involving negative powers of z_0) enter into the commutator. Note also that the last expression in (2.3.13) cannot be expanded by "linearity," since, for instance, the expression $Y(Y(u, z_1 - z_2)v, z_2)$ does not exist (in the usual algebraic sense).

Taking $\mathrm{Res}_{z_0}\mathrm{Res}_{z_1}$ of the Jacobi identity, we find that for $u, v \in V$,

$$[u_0, Y(v, z)] = Y(u_0 v, z) \tag{2.3.14}$$

and in particular,

$$[u_0, v_n] = (u_0 v)_n \quad \text{for} \quad n \in \mathbb{Z}, \tag{2.3.15}$$

$$[u_0, v_0] = (u_0 v)_0; \tag{2.3.16}$$

thus the operators u_0 form a Lie algebra. Note that (2.3.14) generalizes the $L(-1)$-bracket formula (2.3.1) (the case $u = \omega$).

Some important consequences of the axioms involving the operator $L(-1)$ are:

$$e^{z_0 L(-1)} Y(v, z) e^{-z_0 L(-1)} = Y(e^{z_0 L(-1)} v, z) = Y(v, z + z_0) \tag{2.3.17}$$

$$Y(v, z)\mathbf{1} = e^{z L(-1)} v \tag{2.3.18}$$

$$Y(u, z)v = e^{z L(-1)} Y(v, -z)u \quad (\textit{skew-symmetry}) \tag{2.3.19}$$

for $u, v \in V$; (2.3.17) follows from the $L(-1)$-derivative and bracket formulas (2.2.13) and (2.3.1), (2.3.18) then follows by applying to $\mathbf{1}$, and for (2.3.19), we can now use the invariance of each side of the Jacobi identity under the interchange $(u, z_1, z_0) \longleftrightarrow (v, z_2, -z_0)$. (Recall that (2.3.18) has already been observed in (2.2.20).)

2.4. Elementary categorical notions

Given vertex operator algebras V_1, V_2 of the same rank, a *homomorphism* or *map* $f : V_1 \to V_2$ is defined in the obvious way as a grading-preserving linear map such that

$$f(Y(u, z)v) = Y(f(u), z)f(v) \quad \text{for} \quad u, v \in V_1, \tag{2.4.1}$$

or equivalently,

$$f(u_n v) = f(u)_n f(v) \quad \text{for} \quad u, v \in V_1, \quad n \in \mathbb{Z}, \tag{2.4.2}$$

and such that

$$f(\mathbf{1}) = \mathbf{1}, \quad f(\omega) = \omega. \tag{2.4.3}$$

We say that V_1 is a *subalgebra* of V_2 if there is an injective homomorphism $V_1 \to V_2$; in particular, the vacuum vectors and the vectors ω must agree. An *isomorphism* is a bijective homomorphism, an *endomorphism* is a homomorphism from V to itself, and an *automorphism* is a bijective endomorphism. In particular, an isomorphism can be defined as a linear isomorphism f such that

$$f \circ Y(v, z) \circ f^{-1} = Y(f(v), z) \quad \text{for} \quad v \in V_1 \tag{2.4.4}$$

$$f(\omega) = \omega; \tag{2.4.5}$$

then f is grading-preserving and $f(\mathbf{1}) = \mathbf{1}$.

The *subalgebra* $\langle S \rangle$ *generated by a subset* S of a vertex operator algebra V, defined as the smallest subalgebra containing S, is given by:

$$\langle S \rangle = \text{span}\{v^{(1)}_{n_1} \cdots v^{(j)}_{n_j} \cdot v^{(j+1)} | \text{each } v^{(i)} \in \{\text{homogeneous components of elements of } S\} \cup \{\mathbf{1}, \omega\}\}; \tag{2.4.6}$$

in fact, Res_{z_1} of the Jacobi identity shows that for $u, v \in V$ and $m, n \in \mathbb{Z}$, $(u_m v)_n$ can be expressed as a linear combination (finite on any given element of V) of operators of the form $u_p v_q$ and $v_p u_q$. In particular, the smallest subalgebra of V is the Virasoro algebra submodule generated by $\mathbf{1}$:

$$\langle \mathbf{1} \rangle = \text{span}\{L(n_1) \cdots L(n_j) \cdot \mathbf{1} | n_i \in \mathbb{Z}\} \tag{2.4.7}$$

(recall that $\omega = L(-2)\mathbf{1}$.)

Given vertex operator algebras $V_1, V_2, ..., V_n$ of equal rank, the *direct product* vertex operator algebra

$$V = V_1 \times \cdots \times V_n \tag{2.4.8}$$

can be constructed on the direct sum $V_1 \oplus \cdots \oplus V_n$ of the corresponding vector spaces, with grading, vertex operators, $\mathbf{1}$ and ω the natural direct sums of the corresponding things on the V_i. The direct product is characterized by the expected universal property: There are natural projection maps $V \to V_i$, and given a vertex operator algebra U, again of the same rank, and maps $U \to V_i$, there exists a unique map $U \to V$ making the obvious diagram commute. Note that the direct factors V_i are *not* subalgebras of V since the corresponding elements $\mathbf{1}$ and ω do not match. On the other hand, V is a direct sum of submodules V_i for the Virasoro algebra. Observe that the projections of $\mathbf{1}$ in V to the factors V_i are annihilated by $L(-1)$, $L(0)$ and $L(1)$ for V, so that in particular the space of $\mathfrak{sl}(2)$-invariant elements in a vertex operator algebra need not be one-dimensional.

The one-dimensional space $\mathbb{F}\mathbf{1}$ carries a natural vertex operator algebra structure of rank 0 and weight 0 and with $\omega = 0$. We may also think of the space 0 as a degenerate vertex operator algebra of all ranks simultaneously (note that in the Virasoro algebra commutation relations (2.2.9), rank V really refers to a multiple of the identity operator on V); every vertex operator algebra maps to it.

The notions of simple algebra, semisimple algebra, ideal and quotient algebra will be discussed in Section 4.3 below.

2.5. Tensor products

Tensor products of finitely many vertex operator algebras may be constructed: Given $V_1, ..., V_n$, the vector space

$$V = V_1 \otimes \cdots \otimes V_n \tag{2.5.1}$$

becomes a vertex operator algebra of rank equal to the sum of the ranks of the V_i when we provide V with the tensor product grading and we set

$$Y(v_1 \otimes \cdots \otimes v_n, z) = Y(v_1, z) \otimes \cdots \otimes Y(v_n, z) \quad (v_i \in V_i) \tag{2.5.2}$$

$$\mathbf{1} = \mathbf{1} \otimes \cdots \otimes \mathbf{1} \tag{2.5.3}$$

$$\omega = \omega \otimes \mathbf{1} \otimes \cdots \otimes \mathbf{1} + \cdots + \mathbf{1} \otimes \cdots \otimes \mathbf{1} \otimes \omega. \tag{2.5.4}$$

The space V is a tensor product of Virasoro algebra modules. The Jacobi identity for V is most naturally proved using the rationality, commutativity and associativity properties discussed in the next section; see Proposition 3.7.1 below.

Given vertex operator algebras $U_1, ..., U_n$ and homomorphisms $U_i \to V_i$ $(i = 1, ..., n)$, their tensor product is a homomorphism from $U_1 \otimes \cdots \otimes U_n$ to V.

2.6. The Virasoro algebra and primary fields

A weight vector $v \in V_{(h)}$ of weight $h \in \mathbb{Z}$ is called a *lowest weight vector for the Virasoro algebra* if

$$L(n)v = 0 \quad \text{for} \quad n > 0. \tag{2.6.1}$$

It is easy to see from the Jacobi identity and the $L(-1)$-derivative property (2.2.13) that this condition on $v \in V$ is equivalent to the standard condition

$$[L(n), Y(v, z)] = \left(z^{n+1}\frac{d}{dz} + h(n+1)z^n\right) Y(v, z) \quad \text{for} \quad n \in \mathbb{Z} \tag{2.6.2}$$

defining a *primary field of weight h*. Factoring the first-order differential operator on the right-hand side as

$$z^{n+1}\frac{d}{dz} + h(n+1)z^n = z^{n+1}z^{-h(n+1)}\frac{d}{dz}z^{h(n+1)}, \tag{2.6.3}$$

we find by iterating (2.6.2) that (using formal exponential series notation as usual)

$$e^{z_0 L(n)}Y(v, z)e^{-z_0 L(n)} = \left(\frac{z_1}{z}\right)^{h(n+1)} Y(v, z_1) = \left(\frac{\partial z_1}{\partial z}\right)^h Y(v, z_1) \tag{2.6.4}$$

where z_1 is the "formal change of variables"

$$z_1 = e^{z_0 z^{n+1}\frac{d}{dz}}(z) = \begin{cases} e^{z_0}z & \text{if} \quad n = 0 \\ (z^{-n} - nz_0)^{-1/n} & \text{if} \quad n \neq 0; \end{cases} \tag{2.6.5}$$

note that the formal variable z_0 serves as both the "parameter" of the "one-parameter group" indicated in (2.6.5) and also as the "constant of integration" for the differential equation

$$\frac{dz_1}{dz} = \left(\frac{z_1}{z}\right)^{n+1} \tag{2.6.6}$$

satisfied by (2.6.5). In particular, the expression $Y(v, z)$ "transforms as a generalized tensor field" in the standard way.

For $n = 0, \pm 1$, formula (2.6.4) gives

$$e^{z_0 L(n)} Y(v, z) e^{-z_0 L(n)} = (cz + d)^{-2h} Y\left(v, \frac{az + b}{cz + d}\right) \tag{2.6.7}$$

where a, b, c, d are formal series involving z_0 determined by the condition

$$z_1 = \frac{az + b}{cz + d} \tag{2.6.8}$$

(see (2.6.5)). Formula (2.6.7) is equivalent to (2.6.2) for $n = 0, \pm 1$ (the condition that $Y(v, z)$ be a *quasiprimary field of weight* h), and this in turn is equivalent to the condition that the vector $v \in V_{(h)}$ of weight h be a *lowest weight vector for* $\mathfrak{sl}(2)$ in the sense that

$$L(1)v = 0. \tag{2.6.9}$$

2.7. $\mathcal{S}_3$-symmetry of the Jacobi identity

We now derive a symmetry property of the Jacobi identity under the symmetric group $\mathcal{S}_3$, and for later purposes we shall keep track of the properties used in the derivation. Let us retain the axioms for a vertex operator algebra except for the Jacobi identity, and let us call the assertion that (2.2.8) holds when applied to a vector w "the Jacobi identity for the ordered triple (u, v, w)." By skew-symmetry (2.3.19) for the pair (u, v) and the second equality in (2.3.17) for the vector $Y(v, -z_0)u$ we have

$$Y(Y(u, z_0)v, z_2) = Y(e^{z_0 L(-1)} Y(v, -z_0)u, z_2) = Y(Y(v, -z_0)u, z_2 + z_0). \tag{2.7.1}$$

Thus from the general δ-function properties (2.1.11) and (2.1.13), the Jacobi identity for (u, v, w) gives

$$\begin{aligned}(-z_0)^{-1}\delta\left(\frac{z_2 - z_1}{-z_0}\right) Y(v, z_2) Y(u, z_1)w - (-z_0)^{-1}\delta\left(\frac{z_1 - z_2}{-(-z_0)}\right) Y(u, z_1) Y(v, z_2)w \\ = z_1^{-1}\delta\left(\frac{z_2 - (-z_0)}{z_1}\right) Y(Y(v, -z_0)u, z_1)w,\end{aligned} \tag{2.7.2}$$

which is the Jacobi identity for (v, u, w) (with (z_1, z_2, z_0) replaced by $(z_2, z_1, -z_0)$).

On the other hand, multiplying both sides of the Jacobi identity for (u, v, w) by $e^{-z_2L(-1)}$ and using (2.3.19) for the pairs (v, w), $(v, Y(u, z_1)w)$ and $(Y(u, z_0)v, w)$ and the outer equality in (2.3.17) for the vector u, we obtain

$$z_0^{-1}\delta\left(\frac{z_1 - z_2}{z_0}\right)Y(u, z_1 - z_2)Y(w, -z_2)v - z_0^{-1}\delta\left(\frac{z_2 - z_1}{-z_0}\right)Y(Y(u, z_1)w, -z_2)v$$
$$= z_2^{-1}\delta\left(\frac{z_1 - z_0}{z_2}\right)Y(w, -z_2)Y(u, z_0)v. \tag{2.7.3}$$

Then (2.1.11) and (2.1.13) give

$$z_1^{-1}\delta\left(\frac{z_0 + z_2}{z_1}\right)Y(u, z_0)Y(w, -z_2)v + z_2^{-1}\delta\left(\frac{z_0 - z_1}{-z_2}\right)Y(Y(u, z_1)w, -z_2)v$$
$$= z_1^{-1}\delta\left(\frac{-z_2 - z_0}{-z_1}\right)Y(w, -z_2)Y(u, z_0)v, \tag{2.7.4}$$

that is,

$$z_1^{-1}\delta\left(\frac{z_0 - (-z_2)}{z_1}\right)Y(u, z_0)Y(w, -z_2)v$$
$$- z_1^{-1}\delta\left(\frac{(-z_2) - z_0}{-z_1}\right)Y(w, -z_2)Y(u, z_0)v$$
$$= (-z_2)^{-1}\delta\left(\frac{z_0 - z_1}{-z_2}\right)Y(Y(u, z_1)w, -z_2)v, \tag{2.7.5}$$

the Jacobi identity for (u, w, v) (and $(z_0, -z_2, z_1)$). We conclude:

PROPOSITION 2.7.1. *Under the assumptions indicated in the argument above, the Jacobi identity for an ordered triple implies the identity for any permutation of this triple.*

Remark 2.7.2. The second part of this argument also shows that skew-symmetry (2.3.19) for the pair (v, w) and the outer equality in (2.3.17) for the vector u imply that the commutator formula (2.3.13) applied to w (that is, Res_{z_0} of the Jacobi identity for (u, v, w)) is equivalent to the following formula, which, in view of (2.1.11) and (2.1.13), is Res_{z_1} of the Jacobi identity for (u, w, v) :

$$Y(Y(u, z_0)w, z_2)v = Y(u, z_0 + z_2)Y(w, z_2)v$$
$$- \mathrm{Res}_{z_1} z_0^{-1}\delta\left(\frac{z_2 - z_1}{-z_0}\right)Y(w, z_2)Y(u, z_1)v. \tag{2.7.6}$$

(Of course, we replace (z_1, z_2, z_0) by $(z_0, -z_2, z_1)$.) With the help of (2.1.11) and (2.1.14), this last formula can also be written

$$Y(Y(u, z_0)w, z_2)v - Y(u, z_0 + z_2)Y(w, z_2)v$$
$$= Y(w, z_2)(Y(u, z_2 + z_0) - Y(u, z_0 + z_2))v \tag{2.7.7}$$

(cf. (2.3.13)).

2.8. Quasi-vertex operator algebras

In the definition of vertex operator algebra given above, we have assumed the existence of a representation of the Virasoro algebra generated by a vertex operator $Y(\omega, z)$. But in fact most of the results in this paper use only the associated representation of $\mathfrak{sl}(2) = \mathrm{span}(L(-1), L(0), L(1))$. Thus it is perhaps useful to define a notion of "quasi-vertex operator algebra," assuming only the existence and properties of a representation, say ρ, of $\mathfrak{sl}(2)$ on V, and one may observe the validity of most of the results for these objects. We define the *quasi-vertex operator algebra* $(V, Y, \mathbf{1}, \rho)$ to be a $\mathbb{Z}$-graded vector space V equipped with a linear map

$$\begin{aligned} V &\to (\mathrm{End}\, V)[[z, z^{-1}]] \\ v &\mapsto Y(v, z), \end{aligned} \tag{2.8.1}$$

a vacuum vector $\mathbf{1}$ and a representation ρ of $\mathfrak{sl}(2)$ on V given by

$$L(-1) = \rho\left(\begin{pmatrix} 0 & 1 \\ 0 & 0 \end{pmatrix}\right), \quad L(0) = \rho\left(\begin{pmatrix} 1/2 & 0 \\ 0 & -1/2 \end{pmatrix}\right), \quad L(1) = \rho\left(\begin{pmatrix} 0 & 0 \\ -1 & 0 \end{pmatrix}\right), \tag{2.8.2}$$

satisfying all the axioms for a vertex operator algebra except that the axiom (2.2.9), (2.2.10) involving ω is replaced by the bracket formulas (2.3.1)-(2.3.3) and formula (2.3.4) for $n = 0, \pm 1$. There is no notion of rank for a quasi-vertex operator algebra. Homomorphisms, tensor products, etc., are defined in the obvious way. One has the notion of quasiprimary field (recall (2.6.7)-(2.6.9)), and the variable z_1 of (2.6.8) corresponds to the matrix

$$\begin{pmatrix} a & b \\ c & d \end{pmatrix} = e^{z_0 \rho^{-1}(L(n))}, \quad n = 0, \pm 1. \tag{2.8.3}$$

Concepts associated with modules (see Section 4 below) are defined in the natural way.

3. DUALITY FOR VERTEX OPERATOR ALGEBRAS

In this section we shall explain three important properties of a vertex operator algebra – rationality, commutativity and associativity – and we shall show that they may be used in place of the Jacobi identity in the definition of vertex operator algebra. These three properties are aspects of "duality," in physics terminology. After defining the concept of module in Section 4, we shall see that these considerations can also be applied to modules.

3.1. Expansions of rational functions

For our purposes, we shall need a formalism which enables us to expand certain rational functions in both positive and negative integral powers of certain variables. Such expansions correspond, in the complex-variables framework, to different expansions, convergent in different domains, of a common rational function. But in contrast with what is permissible in the complex-variables setting, we may subtract two different expansions of the same rational function to obtain an "expansion of zero" which is nonzero as a formal Laurent series. This point of view is fundamental for us and represents one of the most important features of the formal calculus approach. We have already encountered expansions of zero above; the δ-function $\delta(z_1/z_2)$ is the most basic example, as we shall discuss in more detail below, and expansions of zero pervade the theory of vertex operator algebras, for instance in formulation of the Jacobi identity.

The most important expansions that we shall need are the two opposite expansions of the rational function $(z_1 - z_2)^n$ for n a negative integer. In our arguments, however, we shall naturally encounter rational functions involving linear combinations of more than two formal variables z_i, and so we shall set up our framework in a level of generality which will include such cases. The most general rational functions that we shall require are powers of homogeneous linear expressions. The reader should be aware, however, that the most important cases of the following discussion are the cases $(z_1 \pm z_2)^n$, and keeping these cases in mind will make this discussion easy to follow.

Let S denote the set of nonzero linear polynomials in n variables:

$$S = \left\{ \sum_{i=1}^{n} a_i z_i \middle| a_i \in \mathbb{F},\ a_i \text{ not all zero} \right\} \subset \mathbb{F}[z_1, ..., z_n]. \tag{3.1.1}$$

Consider the subring $\mathbb{F}[z_1, ..., z_n]_S$ of the field of rational functions $\mathbb{F}(z_1, ..., z_n)$ obtained by inverting (localizing with respect to) the products of (zero or more) elements of S. Let $(i_1 \cdots i_n)$ be a permutation of $(1 \cdots n)$. We shall recursively define maps

$$\iota_{i_1 \cdots i_n} : \mathbb{F}[z_1, ..., z_n]_S \to \mathbb{F}[[z_1, z_1^{-1}, ..., z_n, z_n^{-1}]] \tag{3.1.2}$$

as follows: For $n = 1$, ι_1 shall be the inclusion map; in this case,

$$S = \{az_1 | a \in \mathbb{F}^\times\}, \tag{3.1.3}$$

so that

$$\mathbb{F}[z_1]_S = \mathbb{F}[z_1, z_1^{-1}]. \tag{3.1.4}$$

Assume that the maps $\iota_{i_1 \cdots i_{n-1}}$ are defined. To define $\iota_{i_1 \cdots i_n}$, let

$$f(z_1, ..., z_n) \in \mathbb{F}[z_1, ..., z_n]_S, \tag{3.1.5}$$

so that f can be written in the form

$$f(z_1, ..., z_n) = \frac{g(z_1, ..., z_n)}{\prod_{k=1}^r \left(\sum_{j=2}^n a_{kj} z_{i_j}\right) \prod_{\ell=1}^s \left(\sum_{j=1}^n b_{\ell j} z_{i_j}\right)}, \tag{3.1.6}$$

where $g(z_1, ..., z_n) \in \mathbb{F}[z_1, ..., z_n]$, the denominator is nonzero, and $b_{\ell 1} \neq 0$ for $\ell = 1, ..., s$. We can expand

$$\frac{1}{\prod_{\ell=1}^s \left(\sum_{j=1}^n b_{\ell j} z_{i_j}\right)}$$

as a power series in $z_{i_2}, ..., z_{i_n}$ since $b_{\ell 1} \neq 0$. Call this series $h(z_1, ..., z_n)$. Then for each $t \in \mathbb{Z}$ the coefficient of $z_{i_1}^t$ in $g(z_1, ..., z_n)h(z_1, ..., z_n)$ is a polynomial in $z_{i_2}, ..., z_{i_n}$, which we denote by $g_t(z_{i_2}, ..., z_{i_n})$. Using the assumed definition of

$$\iota_{i_2 \cdots i_n} \left(\frac{g_t(z_{i_2}, ..., z_{i_n})}{\prod_{k=1}^r \left(\sum_{j=2}^n a_{kj} z_{i_j}\right)} \right),$$

we set

$$\iota_{i_1 \cdots i_n} f(z_1, ..., z_n) = \sum_{t \in \mathbb{Z}} \iota_{i_2 \cdots i_n} \left(\frac{g_t(z_{i_2}, ..., z_{i_n})}{\prod_{k=1}^r \left(\sum_{j=2}^n a_{kj} z_{i_j}\right)} \right) z_{i_1}^t. \tag{3.1.7}$$

For example, suppose that $n = 2$ and that $\mathbb{F}$ is algebraically closed. Then all nonzero homogeneous polynomials in two variables are inverted,

$$f(z_1, z_2) = \frac{g(z_1, z_2)}{z_{i_2}^r \prod_{\ell=1}^s (b_{\ell 1} z_{i_1} + b_{\ell 2} z_{i_2})}, \tag{3.1.8}$$

and $\iota_{i_1 i_2}$ is the "expansion in negative powers of z_{i_1}" or in "positive powers of z_{i_2}."

The maps $\iota_{i_1 \cdots i_n}$ are clearly injective.

Now suppose that $(i_1 \cdots i_n)$ and $(j_1 \cdots j_n)$ are two permutations of $(1 \cdots n)$. We consider "expansions of zero" in several variables, in the following sense: We set

$$\begin{aligned} \Theta_{i_1 \cdots i_n}^{j_1 \cdots j_n} : & \mathbb{F}[z_1, ..., z_n]_S \to \mathbb{F}[[z_1, z_1^{-1}, ..., z_n, z_n^{-1}]] \\ & f \mapsto \iota_{i_1 \cdots i_n} f - \iota_{j_1 \cdots j_n} f. \end{aligned} \tag{3.1.9}$$

In the case $n = 2$, we shall use the abbreviation

$$\Theta_{i_1 i_2} = \Theta_{i_1 i_2}^{i_2 i_1} : \mathbb{F}[z_1, z_2]_S \to \mathbb{F}[[z_1, z_1^{-1}, z_2, z_2^{-1}]]. \tag{3.1.10}$$

We have

$$\text{Ker } \Theta_{i_1 i_2} = \mathbb{F}[z_1, z_1^{-1}, z_2, z_2^{-1}], \tag{3.1.11}$$

and for $f \in \mathbb{F}[z_1, z_1^{-1}, z_2, z_2^{-1}]$ and $g \in \mathbb{F}[z_1, z_2]_S$,

$$\Theta_{i_1 i_2}(fg) = f \Theta_{i_1 i_2}(g). \tag{3.1.12}$$

Moreover, if $\iota_{12} f = \iota_{21} g$ for $f, g \in \mathbb{F}[z_1, z_2]_S$, then

$$f = g \in \mathbb{F}[z_1, z_1^{-1}, z_2, z_2^{-1}]. \tag{3.1.13}$$

The most important expansion of zero in two variables is

$$\Theta_{12}((z_1 - z_2)^{-1}) = \sum_{i \in \mathbb{Z}} z_1^i z_2^{-1-i} = z_2^{-1} \delta\left(\frac{z_1}{z_2}\right), \tag{3.1.14}$$

(cf. (2.1.11)), an expression which arises naturally in connection with the Jacobi identity. This expression may be heuristically thought of as $\delta_0(z_1 - z_2)$, where δ_0 is the "δ-function at $z = 0$," but we shall not attempt to define δ_0 as a rigorous formal series.

The two δ-function identities (2.1.13) and (2.1.14) have natural and basic generalizations involving rational functions. Notice that in the spirit of the δ-function multiplication principle (2.1.11), the expressions on both sides of (2.1.13) and the three terms occurring in (2.1.14) all correspond to the same formal substitution

$$z_1 = z_0 + z_2; \tag{3.1.15}$$

recall that in (2.1.14), the three expressions are formal power series in z_2, z_1 and z_0, respectively, as has been mentioned after (2.1.14). The following result, proved immediately by simply multiplying (2.1.13) and (2.1.14) by the indicated rational function f and invoking a variant of (2.1.11), may be thought of as a formal algebraic analogue of the Cauchy residue theorem for suitably special meromorphic functions (cf. the Introduction):

PROPOSITION 3.1.1. *Consider a rational function of the form*

$$f(z_0, z_1, z_2) = \frac{g(z_0, z_1, z_2)}{z_0^r z_1^s z_2^t}, \tag{3.1.16}$$

where g is a polynomial and $r, s, t \in \mathbb{Z}$. Then

$$z_1^{-1}\delta\left(\frac{z_2 + z_0}{z_1}\right)\iota_{20}\left(f|_{z_1=z_0+z_2}\right) = z_2^{-1}\delta\left(\frac{z_1 - z_0}{z_2}\right)\iota_{10}\left(f|_{z_2=z_1-z_0}\right) \tag{3.1.17}$$

and

$$z_0^{-1}\delta\left(\frac{z_1 - z_2}{z_0}\right)\iota_{12}\left(f|_{z_0=z_1-z_2}\right) - z_0^{-1}\delta\left(\frac{z_2 - z_1}{-z_0}\right)\iota_{21}\left(f|_{z_0=z_1-z_2}\right)$$
$$= z_2^{-1}\delta\left(\frac{z_1 - z_0}{z_2}\right)\iota_{10}\left(f|_{z_2=z_1-z_0}\right). \tag{3.1.18}$$

3.2. Rationality of products and commutativity

In order to apply this formalism to vertex operator algebras, we consider "matrix coefficients" of products of vertex operators. Set

$$V' = \coprod_{n\in\mathbb{Z}} V^*_{(n)}, \tag{3.2.1}$$

the direct sum of the dual spaces of the homogeneous subspaces $V_{(n)}$ of the vertex operator algebra V – the space of linear functionals on V vanishing on all but finitely many $V_{(n)}$. Denote by $\langle\cdot,\cdot\rangle$ the natural pairing between V' and V. (We shall use these notations for graded vector spaces in general.) Note that for $v, v_1 \in V$ and $v' \in V'$,

$$\langle v', Y(v_1, z_1)v\rangle \in \mathbb{F}[z_1, z_1^{-1}] = \mathbb{F}[z_1]_S, \tag{3.2.2}$$

from (2.3.9). This phenomenon generalizes to the "rationality" property of products and iterates of vertex operators, leading to the "commutativity" and "associativity" properties. We have:

PROPOSITION 3.2.1. *(a)* **(rationality of products)** *For $v, v_1, v_2 \in V$ and $v' \in V'$, the formal series $\langle v', Y(v_1, z_1)Y(v_2, z_2)v\rangle$, which involves only finitely many negative powers of z_2 and only finitely many positive powers of z_1, lies in the image of the map ι_{12} :*

$$\langle v', Y(v_1, z_1)Y(v_2, z_2)v\rangle = \iota_{12} f(z_1, z_2), \tag{3.2.3}$$

where the (uniquely determined) element $f \in \mathbb{F}[z_1, z_2]_S$ is of the form

$$f(z_1, z_2) = \frac{g(z_1, z_2)}{z_1^r z_2^s (z_1 - z_2)^t} \tag{3.2.4}$$

for some $g \in \mathbb{F}[z_1, z_2]$ *and* $r, s, t \in \mathbb{Z}$.

(*b*) **(commutativity)** *We also have*

$$\langle v', Y(v_2, z_2)Y(v_1, z_1)v\rangle = \iota_{21} f(z_1, z_2), \tag{3.2.5}$$

that is,

$$\text{“}Y(v_1, z_1)Y(v_2, z_2) \text{ agrees with } Y(v_2, z_2)Y(v_1, z_1) \tag{3.2.6}$$

as operator-valued rational functions.” In particular,

$$\Theta_{12}(\iota_{12}^{-1}\langle v', Y(v_1, z_1)Y(v_2, z_2)v\rangle) = \langle v', [Y(v_1, z_1), Y(v_2, z_2)]v\rangle. \tag{3.2.7}$$

Proof. By (3.2.2), the expression

$$\langle v', Y((Y(v_1, z_1 - z_2) - Y(v_1, -z_2 + z_1))v_2, z_2)v\rangle \tag{3.2.8}$$

(cf. (2.3.13)) is clearly an expansion of zero of the form $\Theta_{12}(g(z_1, z_2)/z_2^s(z_1 - z_2)^t)$ with $g(z_1, z_2), s, t$ as in (3.2.4). Thus from (2.3.13) we have

$$\begin{aligned} &\langle v', Y(v_1, z_1)Y(v_2, z_2)v\rangle - \iota_{12}\left(\frac{g(z_1, z_2)}{z_2^s(z_1 - z_2)^t}\right) \\ &= \langle v', Y(v_2, z_2)Y(v_1, z_1)v\rangle - \iota_{21}\left(\frac{g(z_1, z_2)}{z_2^s(z_1 - z_2)^t}\right). \end{aligned} \tag{3.2.9}$$

But the left-hand side of (3.2.9) involves only finitely many positive powers of z_1, by (2.3.9), and the right-hand side involves only finitely many negative powers of z_1, by (2.2.5) (or (2.2.16)). Thus each side of (3.2.9) involves only finitely many powers of z_1. On the other hand, the coefficient of each power of z_1 on either side of (3.2.9) is a Laurent polynomial in z_2, by (3.2.2), so that each side of (3.2.9) is of the form $h(z_1, z_2) \in \mathbb{F}[z_1, z_1^{-1}, z_2, z_2^{-1}]$. Then

$$f(z_1, z_2) = \frac{g(z_1, z_2)}{z_2^s(z_1 - z_2)^t} + h(z_1, z_2) \tag{3.2.10}$$

satisfies the desired conditions.

3.3. Rationality of iterates and associativity

It follows from the last result that for $n \in \mathbb{Z}$,

$$\langle v', (z_1 - z_2)^n Y(v_1, z_1)Y(v_2, z_2)v\rangle = \iota_{12}(z_1 - z_2)^n f(z_1, z_2) \tag{3.3.1}$$

and that

$$\begin{aligned} &\iota_{12}^{-1}\left\langle v', z_0^{-1}\delta\left(\frac{z_1 - z_2}{z_0}\right) Y(v_1, z_1)Y(v_2, z_2)v\right\rangle \\ &= \iota_{21}^{-1}\left\langle v', z_0^{-1}\delta\left(\frac{z_2 - z_1}{-z_0}\right) Y(v_2, z_2)Y(v_1, z_1)v\right\rangle, \end{aligned} \tag{3.3.2}$$

$$\Theta_{12}\left(\iota_{12}^{-1}\left\langle v', z_0^{-1}\delta\left(\frac{z_1-z_2}{z_0}\right)Y(v_1,z_1)Y(v_2,z_2)v\right\rangle\right)$$
$$=\left\langle v', \left(z_0^{-1}\delta\left(\frac{z_1-z_2}{z_0}\right)Y(v_1,z_1)Y(v_2,z_2)\right.\right.$$
$$\left.\left.-z_0^{-1}\delta\left(\frac{z_2-z_1}{-z_0}\right)Y(v_2,z_2)Y(v_1,z_1)\right)v\right\rangle \tag{3.3.3}$$

Assuming these formulas, we see that the Jacobi identity asserts:

$$\Theta_{12}\left(\iota_{12}^{-1}\left\langle v', z_0^{-1}\delta\left(\frac{z_1-z_2}{z_0}\right)Y(v_1,z_1)Y(v_2,z_2)v\right\rangle\right)$$
$$=\left\langle v', z_2^{-1}\delta\left(\frac{z_1-z_0}{z_2}\right)Y(Y(v_1,z_0)v_2,z_2)v\right\rangle. \tag{3.3.4}$$

Taking Res_{z_1} of the Jacobi identity and using (2.1.11), (2.1.13) and (2.1.14), we obtain as in (2.7.7) the following analogue of (2.3.13):

$$Y(Y(u,z_0)v,z_2) - Y(u,z_0+z_2)Y(v,z_2)$$
$$= Y(v,z_2)\left(Y(u,z_2+z_0) - Y(u,z_0+z_2)\right). \tag{3.3.5}$$

Just as in Proposition 3.2.1, we use this formula to conclude (admitting 0 into our index set):

PROPOSITION 3.3.1. *(a)* **(rationality of iterates)** *For $v, v_1, v_2 \in V$ and $v' \in V'$, the formal series $\langle v', Y(Y(v_1,z_0)v_2,z_2)v\rangle$, which involves only finitely many negative powers of z_0 and only finitely many positive powers of z_2, lies in the image of the map ι_{20} :*

$$\langle v', Y(Y(v_1,z_0)v_2,z_2)v\rangle = \iota_{20}h(z_0,z_2), \tag{3.3.6}$$

where the (uniquely determined) element $h \in \mathbb{F}[z_0,z_2]_S$ is of the form

$$h(z_0,z_2) = \frac{k(z_0,z_2)}{z_0^r z_2^s (z_0+z_2)^t} \tag{3.3.7}$$

for some $k \in \mathbb{F}[z_0,z_2]$ and $r,s,t \in \mathbb{Z}$.

(b) The series $\langle v', Y(v_1,z_0+z_2)Y(v_2,z_2)v\rangle$, which involves only finitely many negative powers of z_2 and only finitely many positive powers of z_0, lies in the image of ι_{02}, and in fact

$$\langle v', Y(v_1,z_0+z_2)Y(v_2,z_2)v\rangle = \iota_{02}h(z_0,z_2). \tag{3.3.8}$$

That is,

$$\text{“}Y(Y(v_1,z_0)v_2,z_2) \text{ agrees with } Y(v_1,z_0+z_2)Y(v_2,z_2) \tag{3.3.9}$$

as operator-valued rational functions.”

It is also clear that for the rational function $f(z_1,z_2)$ of (3.2.4),

$$\iota_{02}f(z_0+z_2,z_2) = (\iota_{12}f(z_1,z_2))|_{z_1=z_0+z_2}, \tag{3.3.10}$$

so that

$$h(z_0,z_2) = f(z_0+z_2,z_2). \tag{3.3.11}$$

Thus the last two propositions give:

PROPOSITION 3.3.2 (**associativity**). *We have:*

$$\iota_{12}^{-1}\langle v', Y(v_1,z_1)Y(v_2,z_2)v\rangle = \left(\iota_{20}^{-1}\langle v', Y(Y(v_1,z_0)v_2,z_2)v\rangle\right)|_{z_0=z_1-z_2}. \tag{3.3.12}$$

That is,

$$\text{“}Y(v_1,z_1)Y(v_2,z_2) \ \ \textit{agrees with} \ \ Y(Y(v_1,z_1-z_2)v_2,z_2) \tag{3.3.13}$$

as operator-valued rational functions, where the right-hand expression is to be expanded as a Laurent series in $z_1 - z_2$.”

The assertion (3.3.13) is called the “associativity of the operator product expansion” in two-dimensional conformal field theory, and the last result precisely interprets this in an algebraic sense. We emphasize that our algebraic associativity and commutativity results immediately imply the expected results involving convergent series in suitable domains, over a field such as $\mathbb{C}$:

COROLLARY 3.3.3. *Over $\mathbb{C}$, the formal series obtained by taking matrix coefficients of the two expressions in (3.3.13) converge to a common rational function in the domains*

$$|z_1| > |z_2| > 0 \ \ \text{and} \ \ |z_2| > |z_1 - z_2| > 0, \tag{3.3.14}$$

respectively, and in the common domain

$$|z_1| > |z_2| > |z_1 - z_2| > 0, \tag{3.3.15}$$

these two series converge to the common function. Similarly, the formal series obtained by taking matrix coefficients of the two expressions in (3.2.6) converge to a common rational function in the (disjoint) domains

$$|z_1| > |z_2| > 0 \ \ \text{and} \ \ |z_2| > |z_1| > 0, \tag{3.3.16}$$

respectively.

3.4. The Jacobi identity from commutativity and associativity

Using Res_{z_0} and Res_{z_1} of the Jacobi identity – together with (2.3.9), which follows from (2.3.2) and hence from Res_{z_0} of the Jacobi identity – we have established rationality, commutativity and associativity properties of a vertex operator algebra. Conversely, from Proposition 3.1.1 we immediately observe:

PROPOSITION 3.4.1. *The Jacobi identity follows from the rationality of products and iterates, and commutativity and associativity (the assertions of Propositions 3.2.1, 3.3.1 and 3.3.2). In particular, in the definition of the notion of vertex operator algebra, the Jacobi identity may be replaced by these properties.*

Remark 3.4.2. Skew-symmetry (2.3.19) may be recovered analogously. The rationality, commutativity and associativity together imply that

$$\left(\iota_{20}^{-1}\langle v', Y(Y(v_1, z_0)v_2, z_2)v\rangle\right)|_{z_0=z_1-z_2} = f(z_1, z_2)$$
$$= \left(\iota_{10}^{-1}\langle v', Y(Y(v_2, -z_0)v_1, z_1)v\rangle\right)|_{-z_0=z_2-z_1}. \tag{3.4.1}$$

From (3.1.17) and applications of (2.1.13) and (2.1.11), we find by taking Res_{z_1} that

$$Y(Y(v_1, z_0)v_2, z_2) = Y(Y(v_2, -z_0)v_1, z_2 + z_0), \tag{3.4.2}$$

which in conjunction with the $L(-1)$-derivative property (2.2.13) and the faithfulness (2.3.17) gives us skew-symmetry.

3.5. Several variables

The rationality, commutativity and associativity properties and the Jacobi identity extend to several variables, using the ι and Θ maps of (3.1.2) and (3.1.9). For instance, we have the following rationality of products and commutativity in several variables:

PROPOSITION 3.5.1. *For $v_1, v_2, ..., v_n, v \in V$, $v' \in V'$ and any permutation $(i_1 \cdots i_n)$ of $(1 \cdots n)$, the formal series*

$$\langle v', Y(v_{i_1}, z_{i_1})Y(v_{i_2}, z_{i_2}) \cdots Y(v_{i_n}, z_{i_n})v\rangle$$

lies in the image of the map $\iota_{i_1 \cdots i_n}$:

$$\langle v', Y(v_{i_1}, z_{i_1}) \cdots Y(v_{i_n}, z_{i_n})v\rangle = \iota_{i_1 \cdots i_n} f(z_1, ..., z_n), \tag{3.5.1}$$

where the (uniquely determined) element $f \in \mathbb{F}[z_1, ..., z_n]_S$ is independent of the permutation and is of the form

$$f(z_1, ..., z_n) = \frac{g(z_1, ..., z_n)}{\prod_{i=1}^{n} z_i^{r_i} \prod_{j<k}(z_j - z_k)^{s_{jk}}} \tag{3.5.2}$$

for some $g \in \mathbb{F}[z_1, ..., z_n]$ and $r_i, s_{jk} \in \mathbb{Z}$. In particular,

$$\text{“}Y(v_1, z_1) \cdots Y(v_n, z_n) \text{ agrees with } Y(v_{i_1}, z_{i_1}) \cdots Y(v_{i_n}, z_{i_n}) \tag{3.5.3}$$

as operator-valued rational functions.”

Proof (cf. the proof of Proposition 3.2.1). We shall first prove the rationality (the existence, but not yet the permutation-independence, of f). We argue by induction on n. The result is true for $n = 1$ by (3.2.2) and for $n = 2$ by Proposition 3.2.1. For notational convenience, we give the proof only for $n = 3$;

the general argument is the same. Using (2.3.13), the notation (3.1.9) and the rationality for $n=2$, we have:

$$
\begin{aligned}
&\langle v', Y(v_1,z_1)Y(v_2,z_2)Y(v_3,z_3)v\rangle - \langle v', Y(v_2,z_2)Y(v_3,z_3)Y(v_1,z_1)v\rangle\\
&= \langle v', [Y(v_1,z_1), Y(v_2,z_2)Y(v_3,z_3)]v\rangle\\
&= \langle v', [Y(v_1,z_1), Y(v_2,z_2)]Y(v_3,z_3)v\rangle\\
&\quad + \langle v', Y(v_2,z_2)[Y(v_1,z_1), Y(v_3,z_3)]v\rangle\\
&= \langle v', Y((Y(v_1,z_1-z_2) - Y(v_1,-z_2+z_1))v_2,z_2)Y(v_3,z_3)v\rangle\\
&\quad + \langle v', Y(v_2,z_2)Y((Y(v_1,z_1-z_3) - Y(v_1,-z_3+z_1))v_3,z_3)v\rangle\\
&= \Theta_{123}^{231}(g^{(2)}(z_1,z_2,z_3)/z_2^{r_2^{(2)}} z_3^{r_3^{(2)}} (z_1-z_2)^{s_{12}^{(2)}} (z_2-z_3)^{s_{23}^{(2)}})\\
&\quad + \Theta_{123}^{231}(g^{(3)}(z_1,z_2,z_3)/z_2^{r_2^{(3)}} z_3^{r_3^{(3)}} (z_1-z_3)^{s_{13}^{(3)}} (z_2-z_3)^{s_{23}^{(2)}}),
\end{aligned}
\tag{3.5.4}
$$

where $g^{(\ell)} \in \mathbb{F}[z_1,z_2,z_3]$, $r_i^{(\ell)}, s_{jk}^{(\ell)} \in \mathbb{Z}$. Thus

$$
\begin{aligned}
&\langle v', Y(v_1,z_1)Y(v_2,z_2)Y(v_3,z_3)v\rangle\\
&\quad - \iota_{123}(g^{(2)}(z_1,z_2,z_3)/z_2^{r_2^{(2)}} z_3^{r_3^{(2)}} (z_1-z_2)^{s_{12}^{(2)}} (z_2-z_3)^{s_{23}^{(2)}})\\
&\quad - \iota_{123}(g^{(3)}(z_1,z_2,z_3)/z_2^{r_2^{(3)}} z_3^{r_3^{(3)}} (z_1-z_3)^{s_{13}^{(3)}} (z_2-z_3)^{s_{23}^{(2)}})\\
&= \langle v', Y(v_2,z_2)Y(v_3,z_3)Y(v_1,z_1)v\rangle\\
&\quad - \iota_{231}(g^{(2)}(z_1,z_2,z_3)/z_2^{r_2^{(2)}} z_3^{r_3^{(2)}} (z_1-z_2)^{s_{12}^{(2)}} (z_2-z_3)^{s_{23}^{(2)}})\\
&\quad - \iota_{231}(g^{(3)}(z_1,z_2,z_3)/z_2^{r_2^{(3)}} z_3^{r_3^{(3)}} (z_1-z_3)^{s_{13}^{(3)}} (z_2-z_3)^{s_{23}^{(2)}}),
\end{aligned}
\tag{3.5.5}
$$

and since the left-hand side of (3.5.5) involves only finitely many positive powers of z_1 while the right-hand side involves only finitely many negative powers of z_1, we see that each side involves only finitely many powers of z_1. But again by the rationality for $n=2$, the coefficient of each power of z_1 on either side of (3.5.5) is of the form $\iota_{23}(g(z_2,z_3)/z_2^{r_2}z_3^{r_3}(z_2-z_3)^{s_{23}})$, so that each side of (3.5.5) is of the form $\iota_{123}(h(z_1,z_2,z_3)/z_1^{r_1}z_2^{r_2}z_3^{r_3}(z_2-z_3)^{s_{23}})$, where $h \in \mathbb{F}[z_1,z_1^{-1},z_2,z_2^{-1},z_3,z_3^{-1}]$. It follows that the first term on the left-hand side of (3.5.5) is of the desired form, and rationality is proved for $n=3$.

Finally, the permutation-independence (for any n), for transpositions of adjacent indices, follows from the commutator formula (2.3.13) together with the following analogue of (3.1.13): Suppose that rational functions f_1 and f_2 are both of the form (3.5.2), and that $\iota_{1\cdots n}f_1 = \iota_{1\cdots i-1,i+1,i,i+2\cdots n}f_2$ for some i. Then f_1 does not contain the factor $z_i - z_{i+1}$ in the denominator and

$$f_1 = f_2. \tag{3.5.6}$$

Remark 3.5.2. In case $v = \mathbf{1}$, the creation property (2.2.7) shows that the function $f(z_1,\dots,z_n)$ has no pole at $z_i = 0$ $(i=1,\dots,n)$:

$$f(z_1,\dots,z_n) = \frac{g(z_1,\dots,z_n)}{\prod_{j<k}(z_j-z_k)^{s_{jk}}}. \tag{3.5.7}$$

Remark 3.5.3. Note that the assertion of Proposition 3.5.1, even in the special case $v = \mathbf{1}$, implies the assertion for $n-1$ (and arbitrary v), in view of (2.2.7).

Remark 3.5.4. Conversely, suppose that V satisfies all the axioms for a vertex operator algebra except for the Jacobi identity, which we replace by two assumptions: the $L(-1)$-bracket formula (2.3.1) and the $L(0)$-bracket formula (2.3.2) (which implies the homogeneity condition (2.3.9)). Then for $n \geq 3$ and for $n = 1$, the assertion of Proposition 3.5.1 for $n-1$ implies the assertion for n, with $v = \mathbf{1}$. (For $n = 2$, we have the rationality (3.5.1), (3.5.7) but not the commutativity (3.5.3).) Indeed, for $n = 1$ we have

$$\langle v', Y(v_1, z_1)\mathbf{1}\rangle = \langle v', e^{z_1 L(-1)} v_1\rangle, \tag{3.5.8}$$

which is a polynomial in z_1 since $[L(0), L(-1)] = L(-1)$. Now take $n = 3$; the argument here is typical of the general case. We have

$$\begin{aligned}\langle v', Y(v_1, z_1)Y(v_2, z_2)Y(v_3, z_3)\mathbf{1}\rangle &= \langle v', Y(v_1, z_1)Y(v_2, z_2)e^{z_3 L(-1)} v_3\rangle \\ &= \langle v', e^{z_3 L(-1)} Y(v_1, z_1 - z_3)Y(v_2, z_2 - z_3)v_3\rangle,\end{aligned} \tag{3.5.9}$$

which is a finite linear combination of nonnegative powers of z_3 times expressions of the form

$$\langle v'', Y(v_1, z_1 - z_3)Y(v_2, z_2 - z_3)v_3\rangle \tag{3.5.10}$$

for $v'' \in V'$ (recall that V' is characterized as the space of linear functionals on V vanishing on all but finitely many $V_{(n)}$). The rationality for $n = 2$ now implies that (3.5.10) and hence (3.5.9) are of the desired form (3.5.1), (3.5.7). We prove the commutativity assertion for transpositions of adjacent pairs of indices as in the following typical case – that of the transposition of the indices 2 and 3: From the rationality just established,

$$\langle v', Y(v_1, z_1)Y(v_2, z_2)Y(v_3, z_3)\mathbf{1}\rangle = \iota_{123}\left(\frac{g(z_1, z_2, z_3)}{\prod_{j<k}(z_j - z_k)^{s_{jk}}}\right), \tag{3.5.11}$$

$$\langle v', Y(v_1, z_1)Y(v_3, z_3)Y(v_2, z_2)\mathbf{1}\rangle = \iota_{132}\left(\frac{g'(z_1, z_2, z_3)}{\prod_{j<k}(z_j - z_k)^{s'_{jk}}}\right) \tag{3.5.12}$$

as in (3.5.1) and (3.5.7). But in view of (2.3.9), which follows from our assumption (2.3.2), the coefficient of any fixed monomial in z_1 in (3.5.11) is of the form $\langle v'', Y(v_2, z_2)Y(v_3, z_3)\mathbf{1}\rangle$, and analogously for (3.5.12). Now we invoke the assumed commutativity for $n = 2$ to see that the rational functions on the right-hand sides of (3.5.11) and (3.5.12) are equal, as desired.

3.6. The Jacobi identity from commutativity

Now we shall show that rationality and commutativity (for products of two vertex operators), together with the $L(-1)$-bracket formula (2.3.1), the $L(0)$-bracket formula (2.3.2) and the axioms for vertex operator algebras except for the Jacobi identity, imply Proposition 3.3.1. Then as above, associativity (Proposition 3.3.2) and the Jacobi identity will follow. What we shall actually use in the following argument are the rationality and commutativity for products of three vertex operators, with the vector v in Proposition 3.5.1 specialized to $\mathbf{1}$, the $L(-1)$ bracket formula and the axioms except for the Jacobi identity; see Remark 3.5.4.

First note that (2.3.17) and (2.3.18) hold. Let $v_1, v_2, v_3 \in V$ and $v' \in V'$. Then

$$\begin{aligned}
&\langle v', Y(v_3, z_3)Y(Y(v_1, z_0)v_2, z_2)\mathbf{1}\rangle \\
&= \langle v', Y(v_3, z_3)e^{z_2L(-1)}Y(v_1, z_0)v_2\rangle \\
&= \langle v', Y(v_3, z_3)Y(v_1, z_0 + z_2)e^{z_2L(-1)}v_2\rangle \\
&= \langle v', Y(v_3, z_3)Y(v_1, z_0 + z_2)Y(v_2, z_2)\mathbf{1}\rangle, \qquad (3.6.1)
\end{aligned}$$

where of course $Y(v_1, z_0 + z_2)$ is to be expanded in nonnegative integral powers of z_2. We would like to commute $Y(v_3, z_3)$ to the right in the first and last expressions in (3.6.1) (in the sense of rational functions) and then set $z_3 = 0$.

By assumption and Remark 3.5.2, we may write

$$\langle v', Y(v_1, z_1)Y(v_2, z_2)Y(v_3, z_3)\mathbf{1}\rangle = \iota_{123}f(z_1, z_2, z_3), \qquad (3.6.2)$$

where

$$f(z_1, z_2, z_3) = \frac{g(z_1, z_2, z_3)}{(z_1 - z_2)^p(z_1 - z_3)^q(z_2 - z_3)^r} \qquad (3.6.3)$$

and f has the permutation-independence property.

Regarding operations such as ι_{23} for expressions in three variables as acting on each expansion coefficient with respect to z_0, we see using (3.6.1) that as formal series in z_0 with coefficients which are rational functions of z_2 and z_3,

$$\begin{aligned}
&\iota_{23}^{-1}\langle v', Y(Y(v_1, z_0)v_2, z_2)Y(v_3, z_3)\mathbf{1}\rangle \\
&= \iota_{32}^{-1}\langle v', Y(v_3, z_3)Y(Y(v_1, z_0)v_2, z_2)\mathbf{1}\rangle \\
&= \iota_{32}^{-1}\langle v', Y(v_3, z_3)Y(v_1, z_0 + z_2)Y(v_2, z_2)\mathbf{1}\rangle \\
&= \iota_{32}^{-1}\left(\iota_{312}f(z_1, z_2, z_3)|_{z_1=z_0+z_2}\right) \\
&= \iota_{32}^{-1}\left(g(z_1, z_2, z_3)\iota_{12}((z_1 - z_2)^{-p})\iota_{31}((z_1 - z_3)^{-q})\iota_{32}((z_2 - z_3)^{-r})|_{z_1=z_0+z_2}\right) \\
&= \iota_{32}^{-1}\left(g(z_0 + z_2, z_2, z_3)z_0^{-p}\iota_{320}((z_0 + z_2 - z_3)^{-q})\iota_{32}((z_2 - z_3)^{-r})\right) \\
&= g(z_0 + z_2, z_2, z_3)z_0^{-p}\iota_{40}((z_0 + z_4)^{-q})z_4^{-r}|_{z_4=z_2-z_3}, \qquad (3.6.4)
\end{aligned}$$

so that

$$\begin{aligned}
&\langle v', Y(Y(v_1, z_0)v_2, z_2)Y(v_3, z_3)\mathbf{1}\rangle \\
&= \iota_{230}\left(g(z_0 + z_2, z_2, z_3)z_0^{-p}(z_0 + z_2 - z_3)^{-q}(z_2 - z_3)^{-r}\right) \\
&= \iota_{230}(f(z_0 + z_2, z_2, z_3)). \qquad (3.6.5)
\end{aligned}$$

On the other hand,

$$\begin{aligned}
&\langle v', Y(v_1, z_0 + z_2)Y(v_2, z_2)Y(v_3, z_3)\mathbf{1}\rangle \\
&= (\iota_{123} f(z_1, z_2, z_3))\,|_{z_1 = z_0 + z_2} \\
&= \iota_{023}(f(z_0 + z_2, z_2, z_3)).
\end{aligned} \tag{3.6.6}$$

Thus the left-hand sides of (3.6.5) and (3.6.6) are two different expansions of the same rational function. We may set $z_3 = 0$ in both of these expansions and in the common rational function, and we find that

$$\langle v', Y(Y(v_1, z_0)v_2, z_2)v_3\rangle = \iota_{20}(f(z_0 + z_2, z_2, 0)) \tag{3.6.7}$$

$$\langle v', Y(v_1, z_0 + z_2)Y(v_2, z_2)v_3\rangle = \iota_{02}(f(z_0 + z_2, z_2, 0)). \tag{3.6.8}$$

This gives us the assertions of Proposition 3.3.1 and hence 3.3.2:

PROPOSITION 3.6.1. *Rationality of iterates and associativity follow from rationality and commutativity (for products of two vertex operators), the $L(-1)$-bracket formula (2.3.1), the $L(0)$-bracket formula (2.3.2) and the axioms for vertex operator algebras except for the Jacobi identity. In particular, the Jacobi identity may be replaced by commutativity, (2.3.1) and (2.3.2) in the axioms.*

Remark 3.6.2. The argument has shown that we may replace the hypotheses in Proposition 3.6.1 by rationality and commutativity for products of three vertex operators applied to $\mathbf{1}$, (the case $n = 3$, $v = \mathbf{1}$ in Proposition 3.5.1), (2.3.1) and the axioms except for the Jacobi identity. In particular, the Jacobi identity may be replaced by such commutativity and (2.3.1) in the axioms.

3.7. Proof of the tensor product construction

We recall that the Jacobi identity for tensor products of vertex operator algebras was left unproved in Section 2.5. But now either Proposition 3.4.1 or Proposition 3.6.1 immediately yields:

PROPOSITION 3.7.1. *The tensor product of vertex operator algebras, with the structure given by (2.5.1)-(2.5.4), is a vertex operator algebra.*

4. MODULES

4.1. Definition

DEFINITION 4.1.1. Given a vertex operator algebra $(V, Y, \mathbf{1}, \omega)$, a *module for* V (or V*-module* or *representation space*) is a $\mathbb{Q}$-graded vector space (graded by *weights*)

$$W = \coprod_{n \in \mathbb{Q}} W_{(n)}; \quad \text{for} \;\; w \in W_{(n)}, \quad n = \text{wt}\; w; \tag{4.1.1}$$

(cf. (2.2.1)) such that

$$\dim W_{(n)} < \infty \;\; \text{for} \;\; n \in \mathbb{Q}, \tag{4.1.2}$$

$$W_{(n)} = 0 \;\; \text{for} \;\; n \;\; \text{sufficiently small}, \tag{4.1.3}$$

equipped with a linear map $V \otimes W \rightarrow W[[z, z^{-1}]]$, or equivalently,

$$\begin{aligned} V &\rightarrow (\text{End}\; W)[[z, z^{-1}]] \\ v &\mapsto Y(v, z) = \sum_{n \in \mathbb{Z}} v_n z^{-n-1} \quad (\text{where} \;\; v_n \in \text{End}\; W) \end{aligned} \tag{4.1.4}$$

(note that the sum is over $\mathbb{Z}$, not $\mathbb{Q}$), $Y(v, z)$ denoting the *vertex operator associated with* v, such that "all the defining properties of a vertex operator algebra that make sense hold." That is, for $u, v \in V$ and $w \in W$:

$$v_n w = 0 \;\; \text{for} \;\; n \;\; \text{sufficiently large} \tag{4.1.5}$$

(so that $Y(v, z)w \in W((z))$);

$$Y(\mathbf{1}, z) = 1; \tag{4.1.6}$$

$$\begin{aligned} z_0^{-1}\delta\left(\frac{z_1 - z_2}{z_0}\right) Y(u, z_1)Y(v, z_2) - z_0^{-1}\delta\left(\frac{z_2 - z_1}{-z_0}\right) Y(v, z_2)Y(u, z_1) \\ = z_2^{-1}\delta\left(\frac{z_1 - z_0}{z_2}\right) Y(Y(u, z_0)v, z_2) \end{aligned} \tag{4.1.7}$$

(the Jacobi identity for operators on W); note that on the right-hand side, $Y(u, z_0)$ is the operator associated with V; the Virasoro algebra relations hold on W with scalar equal to rank V :

$$[L(m), L(n)] = (m - n)L(m + n) + \frac{1}{12}(m^3 - m)\delta_{m+n,0}(\text{rank}\; V) \tag{4.1.8}$$

for $m, n \in \mathbb{Z}$, where

$$L(n) = \omega_{n+1} \text{ for } n \in \mathbb{Z}, \text{ i.e., } Y(\omega, z) = \sum_{n\in\mathbb{Z}} L(n)z^{-n-2}; \tag{4.1.9}$$

$$L(0)w = nw = (\text{wt } w)w \text{ for } n \in \mathbb{Q} \text{ and } w \in W_{(n)}; \tag{4.1.10}$$

$$\frac{d}{dz}Y(v, z) = Y(L(-1)v, z), \tag{4.1.11}$$

where $L(-1)$ is the operator on V.

This completes the definition. We may denote the module just defined by

$$(W, Y). \tag{4.1.12}$$

Remark 4.1.2. The indexing of the grading by $\mathbb{Q}$ in (4.1.1) is only a convenience. We could use instead an $\mathbb{F}$-grading, with the phrase "n sufficiently small" in (4.1.3) replaced by "n sufficiently small in the sense of modifications by an integer." Note that even for $\mathbb{F} = \mathbb{Q}$, this change would give a different definition allowing more "modules."

Remark 4.1.3. As in the case of vertex operator algebras, the Jacobi identity can be written in terms of component operators (recall Remark 2.2.5 and cf. [FLM]).

Remark 4.1.4. A vertex operator algebra V is clearly a module for itself. As such, it is called the *adjoint module.* We are using the same notation $Y(\cdot, z)$ as for algebras, but no confusion should arise provided V is viewed as the adjoint module.

4.2. Consequences of the definition

As in the case of algebras, we have a number of consequences of the definition:

$$[L(-1), Y(v, z)] = Y(L(-1)v, z) \tag{4.2.1}$$

$$[L(0), Y(v, z)] = Y(L(0)v, z) + zY(L(-1)v, z) \tag{4.2.2}$$

$$[L(1), Y(v, z)] = Y(L(1)v, z) + 2zY(L(0)v, z) + z^2Y(L(-1)v, z). \tag{4.2.3}$$

Formulas (4.1.4), (4.1.10), (4.1.11) and (4.2.2) imply that if $v \in V$ is homogeneous, then

$$\text{wt } v_n = \text{wt } v - n - 1 \tag{4.2.4}$$

as an operator on W. In particular, the operator $x_v(n)$ on W defined by

$$Y(v, z) = \sum_{n\in\mathbb{Z}} x_v(n)z^{-n-\text{wt } v} \tag{4.2.5}$$

when v is a homogeneous vector has weight $-n$:

$$\text{wt } x_v(n) = -n. \tag{4.2.6}$$

As in the case of algebras, the notation $x_v(n)$ may be extended by linearity to arbitrary $v \in V$, and (4.2.6) is valid in general.

Taking Res_{z_0} of the Jacobi identity and using (2.1.14), we obtain the commutator formula

$$\begin{aligned}[Y(u,z_1),Y(v,z_2)] &= \mathrm{Res}_{z_0} z_2^{-1}\delta\left(\frac{z_1-z_0}{z_2}\right) Y(Y(u,z_0)v,z_2)\\ &= Y((Y(u,z_1-z_2)-Y(u,-z_2+z_1))v,z_2).\end{aligned} \tag{4.2.7}$$

As in the case of algebras (recall (2.3.13)), only the singular terms in $Y(u,z_0)v$ enter into the commutator. Formulas (2.3.14)-(2.3.17) hold for modules: For $u, v \in V$,

$$[u_0, Y(v,z)] = Y(u_0v, z) \tag{4.2.8}$$

$$[u_0, v_n] = (u_0v)_n \text{ for } n \in \mathbb{Z}, \tag{4.2.9}$$

$$[u_0, v_0] = (u_0v)_0 \tag{4.2.10}$$

$$e^{z_0L(-1)}Y(v,z)e^{-z_0L(-1)} = Y(e^{z_0L(-1)}v,z) = Y(v,z+z_0). \tag{4.2.11}$$

4.3. Elementary categorical notions

We have the expected algebraic notions: Given modules W_1, W_2 for V, a *homomorphism* or *map* $f : W_1 \to W_2$ is defined as a grading-preserving linear map such that

$$f(Y(v,z)w) = Y(v,z)f(w) \text{ for } v \in V, \ \ w \in W_1. \tag{4.3.1}$$

Isomorphisms (i.e., equivalences), automorphisms, submodules, the submodule generated by a subset, quotient modules, direct sums, irreducible (simple) modules, completely reducible (semisimple) modules, etc., are defined as expected. If $V_1 \to V_2$ is a homomorphism of algebras and W is a V_2-module, then W also becomes a V_1-module. In particular, V_1, V_2 and $V_1 \oplus V_2$ are modules for the direct product algebra $V_1 \times V_2$, $V_1 \oplus V_2$ being the adjoint module.

Remark 4.3.1. Note that a direct product algebra is precisely a vertex operator algebra which is a direct sum of submodules for itself. In fact, if the algebra V can be decomposed as $V_1 \oplus V_2$ as a module, then

$$Y(v_1,z)v_2 = 0 \text{ for } v_i \in V_i, \tag{4.3.2}$$

by skew-symmetry (2.3.19).

Remark 4.3.2. The grading of an irreducible module W is congruent mod $\mathbb{Z}$ to a fixed rational number r :

$$W = \coprod_{n \in \mathbb{Z}+r} W_{(n)}. \tag{4.3.3}$$

A vertex operator algebra is *simple* if it is irreducible as a module for itself, and *semisimple* if it is a direct sum of finitely many irreducible modules for itself, or equivalently, a direct product of finitely many simple algebras.

An *ideal* in a vertex operator algebra V is defined as a submodule of V with respect to the adjoint representation. An ideal W of V is automatically "two-sided" since

$$Y(w, z)v \in W[[z, z^{-1}]] \tag{4.3.4}$$

for $w \in W$, $v \in V$, by skew-symmetry (2.3.19). Using this observation repeatedly, we see that V/W carries a natural vertex operator algebra structure. This *quotient algebra* satisfies the condition

$$\text{rank } (V/W) = \text{rank } V. \tag{4.3.5}$$

4.4. Primary fields

The notions of primary and quasiprimary fields and the results discussed in Section 2.6 all hold for modules, except that the weight of a primary field lies in $\mathbb{Q}$, and the primary or quasiprimary condition on $Y(v, z)$ does not imply anything about the vector v; the module could be 0, for instance.

4.5. Rationality, commutativity, associativity and the Jacobi identity

The entire discussion of rationality, etc., in Sections 3.1-3.5, through Proposition 3.5.1, remains valid for modules, and we have:

PROPOSITION 4.5.1. *The assertions of Propositions 3.2.1, 3.3.1, 3.3.2, 3.4.1 and 3.5.1, and of Corollary 3.3.3, hold for modules. That is, we have rationality of products and iterates, commutativity and associativity, and in the definition of the notion of module, the Jacobi identity may be replaced by rationality, commutativity and associativity.*

In Section 5 below we shall develop further aspects of commutativity for modules, and we shall in particular extend Proposition 3.6.1 to modules.

4.6. Tensor product modules for tensor product algebras

Let $W_1, ..., W_p$ be modules for vertex operator algebras $V_1, ..., V_p$, respectively, and assume that the $\mathbb{Q}$-grading (4.1.1) of each of our modules W is discrete, in that $\{n \in \mathbb{Q} | W_{(n)} \neq 0\}$ lies in $\frac{1}{T}\mathbb{Z}$ for some positive integer T. (This of course holds for irreducible modules.) Then we may construct the tensor product module $W_1 \otimes \cdots \otimes W_p$ for the tensor product algebra $V_1 \otimes \cdots \otimes V_p$, by means of the definition

$$Y(v_1 \otimes \cdots \otimes v_p, z) = Y(v_1, z) \otimes \cdots \otimes Y(v_p, z) \quad (v_i \in V_i). \tag{4.6.1}$$

(The discreteness of the gradings insures that the homogeneous subspaces of $W_1 \otimes \cdots \otimes W_p$ are finite-dimensional.) The tensor product module is in particular a tensor product of Virasoro algebra modules. As in the case of algebras, the Jacobi identity for the tensor product would be awkward to verify directly, but Proposition 4.5.1 immediately gives us the desired result, as in Proposition 3.7.1:

PROPOSITION 4.6.1. *The tensor product of modules for finitely many vertex operator algebras, with the structure given by (4.6.1), is a module for the tensor product algebra.*

Also, given modules $X_1, ..., X_p$ for $V_1, ..., V_p$, respectively, and homomorphisms $W_i \to X_i$ $(i = 1, ..., p)$, their tensor product is a homomorphism from $W_1 \otimes \cdots \otimes W_p$ to $X_1 \otimes \cdots \otimes X_p$.

Remark 4.6.2. It is worth emphasizing that we are considering tensor product modules not for a single algebra, but rather for a tensor product of algebras.

4.7. Irreducibility and tensor products

We shall now discuss how irreducibility behaves under tensor products. So as to avoid the usual subtleties which arise over a non-algebraically-closed field (for instance, as an $\mathbb{R} \otimes \mathbb{R}$-module, $\mathbb{C} \otimes \mathbb{C}$ is a reducible tensor product of irreducible modules), *we shall assume in this discussion that our field $\mathbb{F}$ is algebraically closed* (although some of our considerations will not require this). First we observe:

Remark 4.7.1. For an irreducible module W for a vertex operator algebra V, the commuting ring consists of the scalars, that is, the algebra of all operators on W commuting with the action of the operators induced by V on W consists of just the field $\mathbb{F}$. Indeed, each homogeneous subspace $W_{(n)} (n \in \mathbb{Q})$ is irreducible under the algebra of operators induced by V and preserving $W_{(n)}$, as we see by considering the relation (2.3.9) between the operators $v_m (v \in V,\ m \in \mathbb{Z})$ and the grading of W. We also use that $W_{(n)}$ is finite-dimensional and that $\mathbb{F}$ is algebraically closed.

Using this last fact, we now have:

PROPOSITION 4.7.2. *Given modules $W_1, ..., W_p$ with discrete gradings (see Section 4.6) for vertex operator algebras $V_1, ..., V_p$, respectively, the tensor product $V_1 \otimes \cdots \otimes V_p$-module $W_1 \otimes \cdots \otimes W_p$ is irreducible if and only if $W_1, ..., W_p$ are all irreducible.*

Proof. The "only if" part is trivial. For the "if" part, we observe by choosing all but one v_i in (4.6.1) to be $\mathbf{1}$ that our operators on $W_1 \otimes \cdots \otimes W_p$ include a tensor product of irreducible algebras (for each of which the commuting ring is the scalars) of operators on the W_i. It is now easy to prove using the density theorem that the tensor product is then irreducible (see for example Section 5.8 of [J]).

COROLLARY 4.7.3. *Let $V_1, ..., V_p$ be vertex operator algebras. Then $V_1, ..., V_p$ are all simple if and only if the tensor product algebra $V_1 \otimes \cdots \otimes V_p$ is simple.*

We shall next present a more substantial result – a result whose analogue for, say, (finite-dimensional) full matrix algebras is well known – that (under a mild rationality assumption) the irreducible modules for a tensor product $V_1 \otimes \cdots \otimes V_p$ of vertex operator algebras are precisely the tensor products of irreducible modules for the factors V_i. In preparation, we discuss some general facts. For simplicity of notation, we take $p = 2$, without losing any essential content.

Let W be a $V_1 \otimes V_2$-module, at first not necessarily irreducible. From Res_{z_0} of the Jacobi identity, we see that the actions induced by V_1 $(= V_1 \otimes \mathbf{1})$ and by V_2 $(= \mathbf{1} \otimes V_2)$ on W commute: For $v_i \in V_i$, we have

$$[Y(v_1 \otimes \mathbf{1}, z_1), Y(\mathbf{1} \otimes v_2, z_2)] = 0 \quad \text{on} \quad W \tag{4.7.1}$$

(since $Y(v_1 \otimes \mathbf{1}, z_0)(\mathbf{1} \otimes v_2) \in (V_1 \otimes V_2)[[z_0]]$).

Moreover, taking Res_{z_1} and the constant term in z_0 of the Jacobi identity, we find that

$$\begin{aligned} Y(v_1 \otimes v_2, z_2) &= \mathrm{Res}_{z_0} z_0^{-1} Y(Y(v_1 \otimes \mathbf{1}, z_0)(\mathbf{1} \otimes v_2), z_2) \\ &= \mathrm{Res}_{z_1} (z_1 - z_2)^{-1} Y(v_1 \otimes \mathbf{1}, z_1) Y(\mathbf{1} \otimes v_2, z_2) \\ &\quad - \mathrm{Res}_{z_1} (-z_2 + z_1)^{-1} Y(\mathbf{1} \otimes v_2, z_2) Y(v_1 \otimes \mathbf{1}, z_1), \end{aligned} \tag{4.7.2}$$

so that for all $n \in \mathbb{Z}$,

$(v_1 \otimes v_2)_n$ can be expressed as a linear combination, finite on any given vector, of operators of the form $(v_1 \otimes \mathbf{1})_p (\mathbf{1} \otimes v_2)_q$. (4.7.3)

(Note that we do not need operators of the form $(\mathbf{1} \otimes v_2)_q (v_1 \otimes \mathbf{1})_p$ because of (4.7.1).) It follows that the $V_1 \otimes V_2$-submodule U generated by a given vector $u \in W$ is spanned by elements of the form

$$(v_{1,1} \otimes \mathbf{1})_{m_1} \cdots (v_{1,k} \otimes \mathbf{1})_{m_k} (\mathbf{1} \otimes v_{2,1})_{n_1} \cdots (\mathbf{1} \otimes v_{2,l})_{n_l} u, \tag{4.7.4}$$

where $v_{1,i} \in V_1$ and $v_{2,j} \in V_2$, and $v_{1,i}$, $v_{2,j}$ may be assumed homogeneous. In particular,

$U =$ the "V_1-submodule" generated by the "V_2-submodule" generated by u (and similarly for V_2 and V_1, respectively); (4.7.5)

here and below, when the term "submodule" is used in quotation marks, its meaning is understood to be extended so as to refer just to the operators induced by V_1 or V_2, and not to the existence or finite-dimensionality of weight spaces, etc.

Note that W is similar in spirit to a module for a tensor product of associative algebras.

Set

$$\omega_1 = \omega \otimes \mathbf{1}, \quad \omega_2 = \mathbf{1} \otimes \omega_2, \tag{4.7.6}$$

so that

$$\omega = \omega_1 + \omega_2. \tag{4.7.7}$$

Note that

$$\text{wt } \omega_1 = \text{wt } \omega_2 = 2. \tag{4.7.8}$$

Also set

$$L_i(z) = \sum_{n \in \mathbb{Z}} L_i(n) z^{-n-2} = Y(\omega_i, z) \text{ on } W \quad (i = 1, 2). \tag{4.7.9}$$

Then we have two commuting Virasoro algebras acting on W.

All these comments have obvious analogues for a module for a tensor product $V_1 \otimes \cdots \otimes V_p$ of p algebras. For instance, we have the decomposition

$$L(0) = \sum_{i=1}^{p} L_i(0) \tag{4.7.10}$$

of $L(0)$ into commuting operators.

Now let us assume that the $V_1 \otimes \cdots \otimes V_p$-module W is irreducible. Let $n_0 \in \mathbb{Q}$ be the lowest weight for W (recall Remark 4.3.2), and consider the corresponding (nonzero, finite-dimensional) $L(0)$-eigenspace $W_{(n_0)}$. Since $L_i(0)$ commutes with $L(0)$, it preserves $W_{(n_0)}$, for $i = 1, ..., p$. We now state our result:

THEOREM 4.7.4. *Let W be an irreducible module for the tensor product $V_1 \otimes \cdots \otimes V_p$ of vertex operator algebras, and suppose that for $i = 1, ..., p$, the eigenvalues of the operator $L_i(0)$ on the lowest weight space are rational. Then W is a tensor product of modules, necessarily irreducible, for the V_i.*

Proof. As above, we take $p = 2$ for convenience. Since the lowest weight space $W_{(n_0)}$ is finite-dimensional and $\mathbb{F}$ is algebraically closed, there exists a simultaneous eigenvector

$$w_0 \in W_{(n_0)} \tag{4.7.11}$$

for the commuting operators $L_i(0)$, $i = 1, 2$. Denote by n_1, n_2 the corresponding eigenvalues. Then by hypothesis,

$$n_1, n_2 \in \mathbb{Q}, \tag{4.7.12}$$

and we have

$$n_0 = n_1 + n_2. \tag{4.7.13}$$

Now the $L(-1)$-derivative condition (2.2.13) and the $L(0)$-bracket formula (2.3.2) imply that

$$[L_1(0), Y(v_1 \otimes \mathbf{1}, z)] = Y(L_1(0)(v_1 \otimes \mathbf{1}), z) + z\frac{d}{dz}Y(v_1 \otimes \mathbf{1}, z) \tag{4.7.14}$$

for $v_1 \in V_1$. Thus as in (2.3.9), for homogeneous v_1 and $n \in \mathbb{Z}$,

$$\mathrm{wt}_1(v_1 \otimes \mathbf{1})_n = \mathrm{wt}_1(v_1 \otimes \mathbf{1}) - n - 1, \tag{4.7.15}$$

where wt_1 refers to $L_1(0)$-eigenvalue on both $V_1 \otimes V_2$ and the space of operators on W. In particular, $(v_1 \otimes \mathbf{1})_n$ permutes $L_1(0)$-eigenspaces. Moreover, since $(\mathbf{1} \otimes v_2)_n (v_2 \in V_2)$ commutes with $L_1(0)$, it preserves $L_1(0)$-eigenspaces. Of course, similar statements hold for $L_2(0)$.

By the irreducibility, W is the $V_1 \otimes V_2$-module generated by w_0, and from (4.7.4)-(4.7.5) and the fact that n_0 is the lowest weight for W, we see that W is the direct sum of its simultaneous eigenspaces for $L_1(0)$ and $L_2(0)$ and that the $L_2(0)$-eigenvalues are bounded below by n_2. Analogously, the $L_1(0)$-eigenvalues are bounded below by n_1, and it follows that the lowest weight space $W_{(n_0)}$ is filled up by the simultaneous eigenspace for the operators $L_i(0)$ with eigenvalues n_i. (Of course, the weight spaces of W will usually be direct sums of several simultaneous eigenspaces.)

By the density theorem, the restrictions to $W_{(n_0)}$ of the operators induced by $V_1 \otimes V_2$ and preserving $W_{(n_0)}$ fill up End $W_{(n_0)}$. These operators are the restrictions to $W_{(n_0)}$ of the span of those operators acting on the element u in (4.7.4) such that the $L_1(0)$-weight and $L_2(0)$-weight of the operator is 0. Let A_1 denote the algebra of such restrictions to $W_{(n_0)}$ coming from V_1 (i.e., with $\ell = 0$ in (4.7.4)) and let A_2 denote the algebra of such restrictions coming from V_2 (i.e., with $k = 0$ in (4.7.4)). Then

$$\mathrm{End}\, W_{(n_0)} = A_1 A_2 \tag{4.7.16}$$

and A_1 and A_2 are commuting algebras of operators. Choose an irreducible A_1-submodule X_1 of $W_{(n_0)}$. Then A_1 acts faithfully on X_1 since any element of A_1 which annihilates X_1 annihilates $A_2 \cdot X_1 = A_2 A_1 \cdot X_1 = (\mathrm{End}\, W_{(n_0)}) X_1 = W_{(n_0)}$. Thus A_1 restricts faithfully to End X_1 and hence is isomorphic to a full matrix algebra. Similarly, A_2 is isomorphic to a full matrix algebra. It follows that

$$\mathrm{End}\, W_{(n_0)} = A_1 \otimes A_2. \tag{4.7.17}$$

Then $W_{(n_0)}$ has the structure

$$W_{(n_0)} = X_1 \otimes X_2 \tag{4.7.18}$$

as an irreducible $A_1 \otimes A_2$-module, where X_i is an irreducible A_i-module ($i = 1, 2$); in fact, $X_1 \otimes X_2$ is an irreducbile $A_1 \otimes A_2$-module, and End $W_{(n_0)}$ has only one such.

Let

$$w^0 = x_1 \otimes x_2 \tag{4.7.19}$$

($x_i \in X_i$) be a nonzero decomposable tensor in $W_{(n_0)}$. Let

$$W_i = \text{“}V_i\text{-submodule” of } W \text{ generated by } w^0. \tag{4.7.20}$$

Note that W_i is a V_i-module in the full sense of the definition (since it has rational grading, finite-dimensional weight spaces, etc.) Then W_1 is V_1-irreducible (and similarly for W_2). In fact, consideration of the grading shows that any nonzero "V_1-submodule" of W_1 not intersecting $W_{(n_0)}$ would give rise to a nonzero $V_1 \otimes V_2$-submodule of W not intersecting $W_{(n_0)}$. Thus any nonzero "V_1-submodule" of W_1 must intersect $W_{(n_0)}$. But the irreducible A_1-module $A_1 \cdot w^0$ is the full intersection of W_1 and $W_{(n_0)}$, so that the "V_1-submodule" must contain w^0 and hence be all of W_1. This proves the V_1-irreducibility of W_1.

Finally, to show that W is isomorphic to $W_1 \otimes W_2$, consider the abstract tensor product $V_1 \otimes V_2$-module $W_1 \otimes W_2$, where W_i is the V_i-module defined in (4.7.20). Define a linear map

$$\begin{aligned} \varphi : \quad & W_1 \otimes W_2 \to W \\ a_1 \cdot w^0 \otimes a_2 \cdot w^0 & \mapsto a_1 a_2 \cdot w^0, \end{aligned} \tag{4.7.21}$$

where a_i is any operator induced by V_i. Then φ is well defined and is a $V_1 \otimes V_2$-module map. Since $W_1 \otimes W_2$ is irreducible by Proposition 4.7.2, φ is an isomorphism.

5. DUALITY FOR MODULES

In Proposition 4.5.1 we have extended the simplest aspects of the "duality" concepts (rationality, commutativity and associativity) to modules. Here we shall develop these ideas more deeply.

5.1. Duality for one module element and two algebra elements

For vertex operator algebras, rationality and commutativity for products of three vertex operators applied to $\mathbf{1}$, the $L(-1)$-bracket formula (2.3.1) and the axioms except for the Jacobi identity imply the rationality of iterates and associativity, and hence the Jacobi identity (see Remark 3.6.2). However, the rationality of products and commutativity for the vertex operator $Y(v, z)$ in the definition of module will not be enough to insure the rationality of iterates and associativity for modules. Here we shall present a motivational result – the analogue of Remark 3.6.2 for modules, then the main definition (5.1.5) of vertex operators parametrized by module elements, then the basic theory of these operators, and finally, the appropriate analogue of Proposition 3.6.1.

Let V be a vertex operator algebra and let W be a space satisfying all the conditions for a V-module except for the Jacobi identity. The argument proving the assertion of Remark 3.6.2 suggests that we assume the existence of a linear map $W \otimes V \to W[[z, z^{-1}]]$, or equivalently,

$$\begin{aligned} W &\to (\mathrm{Hom}(V, W))[[z, z^{-1}]] \\ w &\mapsto Y(w, z), \end{aligned} \tag{5.1.1}$$

such that

$$Y(w, z)\mathbf{1} = w. \tag{5.1.2}$$

We also assume the following version of rationality and commutativity for products of three operators applied to $\mathbf{1}$: For $u_1, u_2 \in V$, $u_3 \in W$, $w' \in W'$ (defined as in (3.2.1)) and any permutation $(i_1 i_2 i_3)$ of (123), the formal series $\langle w', Y(u_{i_1}, z_{i_1}) Y(u_{i_2}, z_{i_2}) Y(u_{i_3}, z_{i_3})\mathbf{1}\rangle$ lies in the image of the map $\iota_{i_1 i_2 i_3}$:

$$\langle w', Y(u_{i_1}, z_{i_1}) Y(u_{i_2}, z_{i_2}) Y(u_{i_3}, z_{i_3})\mathbf{1}\rangle = \iota_{i_1 i_2 i_3} f(z_1, z_2, z_3), \tag{5.1.3}$$

where the (uniquely determined) element $f \in \mathbb{F}[z_1, z_2, z_3]_S$ is independent of the permutation and is of the form

$$f(z_1, z_2, z_3) = \frac{g(z_1, z_2, z_3)}{(z_1 - z_2)^p (z_1 - z_3)^q (z_2 - z_3)^r} \tag{5.1.4}$$

for some $g \in \mathbb{F}[z_1, z_2, z_3]$ and $p, q, r \in \mathbb{Z}$. (Note that the operators $Y(u_i, z_i)$ are of different types. Note also the degenerate cases u_1 or $u_2 = \mathbf{1}$.) Then the proof of Proposition 3.6.1 (or rather, the proof of the assertion of Remark 3.6.2) gives:

PROPOSITION 5.1.1. *Under the stated conditions, the rationality of iterates and associativity, and in particular, the Jacobi identity, hold for W. That is, W is a V-module.*

(Note that the $L(-1)$-bracket formula (2.3.1) does not have to be assumed for $Y(u_3, z)$.)

Recalling from Section 2.7 the notion of the Jacobi identity for an ordered triple of elements, we shall call the rationality, commutativity and associativity assertions involving the expressions $Y(v_1, z_1)$, $Y(v_2, z_2)$ and $Y(Y(v_1, z_0)v_2, z_2)$ acting on an element v_3 the "rationality, commutativity and associativity for the ordered triple (v_1, v_2, v_3)." Similarly we have the notion of rationality and commutativity for $(v_1, ..., v_n, v_{n+1})$. The arguments in Section 3 show that rationality, commutativity and associativity for a particular triple (v_1, v_2, v_3) are equivalent to the Jacobi identity for (v_1, v_2, v_3).

Now we proceed to define vertex operators parametrized by module elements. Let W be a V-module. In view of the symmetry of the commutativity hypothesis expressed in (5.1.3)-(5.1.4), and also the symmetry of the Jacobi identity presented in Proposition 2.7.1, it is natural to expect the rationality of iterates and associativity, and in particular, the Jacobi identity, to hold for all permutations of the elements u_1, u_2, u_3 – not just for the ordered triple (u_1, u_2, u_3). (Recall that $u_1, u_2 \in V$ and $u_3 \in W$.) If this is to be the case, then we are required to define the operator $Y(w, z)$ of (5.1.1) by the skew-symmetry condition

$$Y(w, z)v = e^{zL(-1)}Y(v, -z) \quad \text{for } w \in W, \ \ v \in V \tag{5.1.5}$$

(recall (2.3.19)). Note that the creation property (5.1.2) indeed holds for this operator. Moreover, the assertion of (4.2.11) holds for $Y(w, z)$, as (4.2.11) itself implies:

$$e^{z_0L(-1)}Y(w, z)e^{-z_0L(-1)}v = Y(e^{z_0L(-1)}w, z)v = Y(w, z + z_0)v \tag{5.1.6}$$

(cf. (2.3.17)). We shall see in addition that the operator (5.1.5) also satisfies the commutativity condition given in (5.1.3)-(5.1.4). In fact, we have:

PROPOSITION 5.1.2. *Let W be a V-module. Define the operator $Y(w, z)$ for $w \in W$ by the skew-symmetry condition (5.1.5), so that (5.1.1), (5.1.2) and (5.1.6) hold. Let $u_1, u_2 \in V$ and $u_3 \in W$. Then for every permutation of (u_1, u_2, u_3), the Jacobi identity and in particular, rationality, commutativity, associativity and the analytic assertions of Corollary 3.3.3, hold. Moreover, for $v_1, ..., v_n \in V$ and $v_{n+1} \in W$, rationality and commutativity hold for $(v_{i_1}, ..., v_{i_{n+1}})$, where $(i_1 \cdots i_{n+1})$ is a permutation of $(1 \cdots n+1)$, and in particular, rationality and commutativity hold for $(u_1, u_2, u_3, \mathbf{1})$.*

Proof. Since W is a module, we have the Jacobi identity for (u_1, u_2, u_3). Then using the definition of $Y(w, z)$ and the proof of Proposition 2.7.1 we get the Jacobi identity for all permutations of (u_1, u_2, u_3). Using these identities and the proofs of Propositions 3.2.1, 3.3.1, 3.3.2, and 3.5.1 and of Corollary 3.3.3, we obtain the rationality, commutativity and associativity assertions.

We have another converse of Proposition 5.1.2, besides those presented in Propositions 4.5.1 and 5.1.1 – a symmetrical analogue of Proposition 3.4.1:

PROPOSITION 5.1.3. *Assume that (W, Y) satisfies all the conditions for a V-module except for the Jacobi identity, and assume the $L(-1)$-bracket formula (4.2.1) for $Y(v, z)$ acting on W. Define $Y(w, z)$ for $w \in W$ by (5.1.5). Fix a permutation $(i_1 i_2 i_3)$ of (123). If rationality, commutativity and associativity hold for $(u_{i_1}, u_{i_2}, u_{i_3})$, where $u_1, u_2 \in V$ and $u_3 \in W$, then (W, Y) is a V-module.*

Proof. Using the assumed rationality, commutativity and associativity and the proof of Proposition 3.4.1 we obtain the Jacobi identity for $(u_{i_1}, u_{i_2}, u_{i_3})$. But (4.2.11) and (5.1.6) hold, and the proof of Proposition 2.7.1 now gives the Jacobi identity for (u_1, u_2, u_3). Thus (W, Y) is a module.

Remark 5.1.4. In the setting of Proposition 5.1.3, the assertion of Proposition 2.7.1 holds – the S_3-symmetry of the Jacobi identity – for permutations of (u_1, u_2, u_3).

Finally, from Proposition 5.1.1 and the argument in Remark 3.5.4, we can give still another criterion for insuring that we have a module:

PROPOSITION 5.1.5. *Assume the conditions of Proposition 5.1.3, and define $Y(w, z)$ for $w \in W$ by (5.1.5). Assume also the $L(0)$-bracket formula (4.2.2) for $Y(v, z)$ $(v \in V)$ acting on W. If rationality and commutativity hold for $(u_{i_1}, u_{i_2}, u_{i_3})$, where $u_1, u_2 \in V$ and $u_3 \in W$, and $(i_1 i_2 i_3)$ is an arbitrary permutation of (123), then (W, Y) is a V-module.*

(Note that we do not need to assume the $L(0)$-bracket formula (2.3.2) for $Y(u_3, z)$.)

5.2. Adjoint vertex operators and the contragredient module

In order to extend the discussion of the previous section to duality for two module elements, we shall need the theory of adjoint operators.

Let (W, Y), with

$$W = \coprod_{n \in \mathbb{Q}} W_{(n)}, \tag{5.2.1}$$

be a module for a vertex operator algebra $(V, Y, \mathbf{1}, \omega)$,

$$W' = \coprod_{n \in \mathbb{Q}} W^*_{(n)} \tag{5.2.2}$$

the graded dual space of W (as in (3.2.1)). We define the *adjoint vertex operators* $Y'(v, z)$ $(v \in V)$ by means of the linear map

$$\begin{aligned} V &\to (\operatorname{End} W')[[z, z^{-1}]] \\ v &\mapsto Y'(v, z) = \sum_{n \in \mathbb{Z}} v'_n z^{-n-1} \quad (\text{where } v'_n \in \operatorname{End} W'), \end{aligned} \tag{5.2.3}$$

determined by the condition

$$\langle Y'(v, z)w', w\rangle = \langle w', Y(e^{zL(1)}(-z^{-2})^{L(0)}v, z^{-1})w\rangle \tag{5.2.4}$$

for $v \in V$, $w' \in W'$, $w \in W$. (Of course, the word "adjoint" has a different sense here from that in Remark 4.1.4.) The operator $(-z^{-2})^{L(0)}$ has the obvious meaning; it acts on a vector of weight $n \in \mathbb{Z}$ as multiplication by $(-z^{-2})^n$. Also note that $e^{zL(1)}(-z^{-2})^{L(0)}v$ involves only finitely many (integral) powers of z, that the right-hand side of (5.2.4) is a Laurent polynomial in z, and that in view of (4.2.4), the components v'_n of the formal Laurent series $Y'(v, z)$ defined by (5.2.4) indeed preserve W'.

The next proposition will serve to motivate the definition of adjoint operator. The Lie algebra $\mathfrak{sl}(2)$ spanned by $L(-1)$, $L(0)$, $L(1)$ will play an important role in the proof below, especially through formal relations among exponentials of these operators. Such exponentials should be thought of in terms of representations of the Lie group $SL(2, \mathbb{C})$, but we shall not need to use this group for anything more than motivation for our algebraic arguments.

We give the space W' a $\mathbb{Q}$-grading by setting

$$W'_{(n)} = W^*_{(n)} \quad \text{for } n \in \mathbb{Q}. \tag{5.2.5}$$

The following proposition defines the *V-module contragredient to W* (we use the term "contragredient" rather than "dual" to avoid confusion with the concept of "duality" discussed in this work):

THEOREM 5.2.1. *The pair (W', Y') carries the structure of a V-module.*

Proof. Conditions (4.1.1)-(4.1.6) are clear. We shall prove the Jacobi identity (4.1.7) and the $L(-1)$-derivative property (4.1.11) below. For the Virasoro algebra properties, we note that

$$\langle Y'(\omega, z)w', w\rangle = \langle w', Y(z^{-4}\omega, z^{-1})w\rangle \tag{5.2.6}$$

since

$$L(1)\omega = 0 \tag{5.2.7}$$

by (2.3.12). Thus, defining component operators $L'(n)$ by

$$Y'(\omega, z) = \sum_{n\in\mathbb{Z}} L'(n)z^{-n-2}, \tag{5.2.8}$$

we have

$$\begin{aligned}\left\langle \sum_{n\in\mathbb{Z}} L'(n)z^{-n}w', w\right\rangle &= \langle z^2 Y'(\omega, z)w', w\rangle \\ &= \langle w', z^{-2}Y(\omega, z^{-1})w\rangle \\ &= \left\langle w', \sum_{n\in\mathbb{Z}} L(-n)z^{-n}w\right\rangle,\end{aligned} \tag{5.2.9}$$

and so

$$\langle L'(n)w', w\rangle = \langle w', L(-n)w\rangle \quad \text{for } n \in \mathbb{Z}. \tag{5.2.10}$$

This immediately gives us (4.1.8)-(4.1.10).

Only properties (4.1.7) and (4.1.11) remain. For these, we shall use some commutator formulas motivated by the Lie group $SL(2, \mathbb{C})$, but formulated and proved in terms of formal series, as usual.

LEMMA 5.2.2. *Let*

$$f(z) \in z\mathbb{F}[[z]]. \tag{5.2.11}$$

We have the following identities, valid on any module for the Lie algebra $\mathfrak{sl}(2)$ *spanned by* $L(-1)$, $L(0)$, $L(1)$:

$$L(-1)e^{f(z)L(0)} = e^{f(z)L(0)}L(-1)e^{-f(z)} \tag{5.2.12}$$

$$L(1)e^{f(z)L(0)} = e^{f(z)L(0)}L(1)e^{f(z)} \tag{5.2.13}$$

$$\begin{aligned} &L(-1)e^{f(z)L(1)} \\ &= e^{f(z)L(1)}L(-1) - 2f(z)L(0)e^{f(z)L(1)} - f(z)^2 L(1)e^{f(z)L(1)} \\ &= e^{f(z)L(1)}L(-1) - 2f(z)e^{f(z)L(1)}L(0) + f(z)^2 e^{f(z)L(1)}L(1) \end{aligned} \tag{5.2.14}$$

These identities also hold for more general f *for which the series are well defined, such as*

$$f(z, z_0) \in z\mathbb{F}[[z, z_0]]. \tag{5.2.15}$$

Proof. From the bracket formula

$$[L(0), L(-1)] = L(-1) \tag{5.2.16}$$

we get

$$e^{f(z)L(0)}L(-1)e^{-f(z)L(0)} = e^{f(z)}L(-1), \tag{5.2.17}$$

which gives (5.2.12). The proof of (5.2.13) is similar.

To prove (5.2.14) we use the bracket relations for $L(-1)$, $L(0)$ and $L(1)$ and induction on n to verify that

$$L(1)^n L(0) = (L(0) + n)L(1)^n \tag{5.2.18}$$

and with the help of this formula, that

$$L(-1)L(1)^n = L(1)^n L(-1) - 2nL(0)L(1)^{n-1} - n(n-1)L(1)^{n-1}, \tag{5.2.19}$$

$$L(-1)L(1)^n = L(1)^n L(-1) - 2nL(1)^{n-1}L(0) + n(n-1)L(1)^{n-1}. \tag{5.2.20}$$

Using (5.2.19) we have

$$
\begin{aligned}
L(-1)e^{f(z)L(1)} &= L(-1)\sum_{n\geq 0}\frac{f(z)^n L(1)^n}{n!}\\
&= \sum_{n\geq 0}\frac{f(z)^n L(-1)L(1)^n}{n!}\\
&= \sum_{n\geq 0}\frac{f(z)^n}{n!}\left(L(1)^n L(-1) - 2nL(0)L(1)^{n-1} - n(n-1)L(1)^{n-1}\right)\\
&= \sum_{n\geq 0}\frac{f(z)^n L(1)^n}{n!}L(-1) - 2f(z)L(0)\sum_{n\geq 1}\frac{f(z)^{n-1}L(1)^{n-1}}{(n-1)!}\\
&\quad - f(z)^2 L(1)\sum_{n\geq 2}\frac{f(z)^{n-2}L(1)^{n-2}}{(n-2)!}\\
&= e^{f(z)L(1)}L(-1) - 2f(z)L(0)e^{f(z)L(1)} - f(z)^2 L(1)e^{f(z)L(1)},
\end{aligned}
$$

and we prove the second equality in (5.2.14) similarly, using (5.2.20).

Now we establish the $L(-1)$-derivative property (4.1.11). For convenience, we assume that $v \in V$ is homogeneous of weight $n \in \mathbb{Z}$:

$$L(0)v = nv. \tag{5.2.21}$$

Using the definition (5.2.4) of $Y(\cdot, z)$ and the chain rule we get

$$
\begin{aligned}
\langle \frac{d}{dz}Y'(v,z)w', w\rangle &= \frac{d}{dz}\langle w', Y(e^{zL(1)}(-z^{-2})^{L(0)}v, z^{-1})w\rangle\\
&= \langle w', \frac{d}{dz}Y(e^{zL(1)}(-z^{-2})^{L(0)}v, z^{-1})w\rangle\\
&= \langle w', Y(\frac{d}{dz}(e^{zL(1)}(-z^{-2})^{L(0)})v, z^{-1})w\rangle\\
&\quad + \langle w', \frac{d}{dz}Y(v_1, z^{-1})\big|_{v_1 = e^{zL(1)}(-z^{-2})^{L(0)}v} w\rangle,
\end{aligned} \tag{5.2.22}
$$

where w' and w are arbitrary elements of W' and W, respectively. We perform the indicated calculations:

$$
\begin{aligned}
&\frac{d}{dz}(e^{zL(1)}(-z^{-2})^{L(0)})\\
&= L(1)e^{zL(1)}(-z^{-2})^{L(0)} - 2z^{-1}e^{zL(1)}L(0)(-z^{-2})^{L(0)}
\end{aligned} \tag{5.2.23}
$$

$$
\begin{aligned}
&\frac{d}{dz}Y(v_1, z^{-1})\big|_{v_1=e^{zL(1)}(-z^{-2})^{L(0)}v}\\
&= -z^{-2}\frac{d}{dz^{-1}}Y(v_1, z^{-1})\big|_{v_1=e^{zL(1)}(-z^{-2})^{L(0)}v}\\
&= -z^{-2}Y(L(-1)v_1, z^{-1})\big|_{v_1=e^{zL(1)}(-z^{-2})^{L(0)}v}\\
&= -z^{-2}Y(L(-1)e^{zL(1)}(-z^{-2})^{L(0)}v, z^{-1})\\
&= -z^{-2}Y((e^{zL(1)}L(-1) - 2ze^{zL(1)}L(0)\\
&\qquad + z^2L(1)e^{zL(1)})(-z^{-2})^n v, z^{-1})\\
&= Y(e^{zL(1)}(-z^{-2})^{n+1}L(-1)v, z^{-1})\\
&\qquad + Y(2z^{-1}e^{zL(1)}L(0)(-z^{-2})^n v, z^{-1})\\
&\qquad - Y(L(1)e^{zL(1)}(-z^{-2})^n v, z^{-1})\\
&= Y(e^{zL(1)}(-z^{-2})^{L(0)}L(-1)v, z^{-1})\\
&\qquad + Y(2z^{-1}e^{zL(1)}L(0)(-z^{-2})^{L(0)}v, z^{-1})\\
&\qquad - Y(L(1)e^{zL(1)}(-z^{-2})^{L(0)}v, z^{-1}).
\end{aligned}
\tag{5.2.24}
$$

Here we have used the outer equality in (5.2.14) and the fact that

$$
L(0)L(-1)v = L(-1)(L(0)+1)v = (n+1)L(-1)v. \tag{5.2.25}
$$

Substituting (5.2.23) and (5.2.24) into (5.2.22) we get

$$
\begin{aligned}
&\langle \frac{d}{dz}Y'(v, z)w', w\rangle\\
&= \langle w', Y(L(1)e^{zL(1)}(-z^{-2})^{L(0)}v\\
&\qquad - 2z^{-1}e^{zL(1)}L(0)(-z^{-2})^{L(0)}v, z^{-1})w\rangle\\
&\quad + \langle w', Y(e^{zL(1)}(-z^{-2})^{L(0)}L(-1)v, z^{-1})w\rangle\\
&\quad + \langle w', Y(2z^{-1}e^{zL(1)}L(0)(-z^{-2})^{L(0)}v, z^{-1})w\rangle\\
&\quad - \langle w', Y(L(1)e^{zL(1)}(-z^{-2})^{L(0)}v, z^{-1})w\rangle\\
&= \langle w', Y(e^{zL(1)}(-z^{-2})^{L(0)}L(-1)v, z^{-1})w\rangle\\
&= \langle Y'(L(-1)v, z)w', w\rangle,
\end{aligned}
\tag{5.2.26}
$$

proving (4.1.11).

Finally, we shall prove the Jacobi identity (4.1.7). Let $v_1, v_2 \in V$, $w \in W$ and $w' \in W'$. What we want to prove can be written as follows:

$$
\begin{aligned}
&\langle z_0^{-1}\delta\left(\frac{z_1 - z_2}{z_0}\right)Y'(v_1, z_1)Y'(v_2, z_2)w', w\rangle\\
&\quad - \langle z_0^{-1}\delta\left(\frac{z_2 - z_1}{-z_0}\right)Y'(v_2, z_2)Y'(v_1, z_1)w', w\rangle\\
&= \langle z_2^{-1}\delta\left(\frac{z_1 - z_0}{z_2}\right)Y'(Y(v_1, z_0)v_2, z_2)w', w\rangle.
\end{aligned}
\tag{5.2.27}
$$

But by the definition (5.2.4) of adjoint vertex operator we have

$$\begin{aligned}&\langle Y'(v_1, z_1)Y'(v_2, z_2)w', w\rangle \\ &= \langle w', Y(e^{z_2L(1)}(-z_2^{-2})^{L(0)}v_2, z_2^{-1})Y(e^{z_1L(1)}(-z_1^{-2})^{L(0)}v_1, z_1^{-1})w\rangle\end{aligned} \tag{5.2.28}$$

$$\begin{aligned}&\langle Y'(v_2, z_2)Y'(v_1, z_1)w', w\rangle \\ &= \langle w', Y(e^{z_1L(1)}(-z_1^{-2})^{L(0)}v_1, z_1^{-1})Y(e^{z_2L(1)}(-z_2^{-2})^{L(0)}v_2, z_2^{-1})w\rangle\end{aligned} \tag{5.2.29}$$

$$\begin{aligned}&\langle Y'(Y(v_1, z_0)v_2, z_2)w', w\rangle \\ &= \langle w', Y(e^{z_2L(1)}(-z_2^{-2})^{L(0)}Y(v_1, z_0)v_2, z_2^{-1})w\rangle,\end{aligned} \tag{5.2.30}$$

and from the Jacobi identity for W we have

$$\begin{aligned}&\langle w', \left(\frac{-z_0}{z_1z_2}\right)^{-1}\delta\left(\frac{z_1^{-1}-z_2^{-1}}{-z_0/z_1z_2}\right)Y(e^{z_1L(1)}(-z_1^{-2})^{L(0)}v_1, z_1^{-1})\cdot \\ &\qquad\cdot Y(e^{z_2L(1)}(-z_2^{-2})^{L(0)}v_2, z_2^{-1})w\rangle \\ &- \langle w', \left(\frac{-z_0}{z_1z_2}\right)^{-1}\delta\left(\frac{z_2^{-1}-z_1^{-1}}{z_0/z_1z_2}\right)Y(e^{z_2L(1)}(-z_2^{-2})^{L(0)}v_2, z_2^{-1})\cdot \\ &\qquad\cdot Y(e^{z_1L(1)}(-z_1^{-2})^{L(0)}v_1, z_1^{-1})w\rangle \\ &= \langle w', (z_2^{-1})^{-1}\delta\left(\frac{z_1^{-1}+z_0/z_1z_2}{z_2^{-1}}\right)\cdot \\ &\qquad\cdot Y(Y(e^{z_1L(1)}(-z_1^{-2})^{L(0)}v_1, -z_0/z_1z_2)e^{z_2L(1)}(-z_2^{-2})^{L(0)}v_2, z_2^{-1})w\rangle,\end{aligned} \tag{5.2.31}$$

or equivalently,

$$\begin{aligned}&- \langle w', z_0^{-1}\delta\left(\frac{z_2-z_1}{-z_0}\right)Y(e^{z_1L(1)}(-z_1^{-2})^{L(0)}v_1, z_1^{-1})\cdot \\ &\qquad\cdot Y(e^{z_2L(1)}(-z_2^{-2})^{L(0)}v_2, z_2^{-1})w\rangle \\ &+ \langle w', z_0^{-1}\delta\left(\frac{z_1-z_2}{z_0}\right)Y(e^{z_2L(1)}(-z_2^{-2})^{L(0)}v_2, z_2^{-1})\cdot \\ &\qquad\cdot Y(e^{z_1L(1)}(-z_1^{-2})^{L(0)}v_1, z_1^{-1})w\rangle \\ &= \langle w', z_1^{-1}\delta\left(\frac{z_2+z_0}{z_1}\right)\cdot \\ &\qquad\cdot Y(Y(e^{z_1L(1)}(-z_1^{-2})^{L(0)}v_1, -z_0/z_1z_2)e^{z_2L(1)}(-z_2^{-2})^{L(0)}v_2, z_2^{-1})w\rangle.\end{aligned} \tag{5.2.32}$$

(As usual, the reader should be observing that the formal Laurent series which arise are well defined.) Thus (by (2.1.13)) the desired result (5.2.27) is equivalent to:

$$\begin{aligned} & z_1^{-1}\delta\left(\frac{z_2+z_0}{z_1}\right) Y(e^{z_2L(1)}(-z_2^{-2})^{L(0)}Y(v_1, z_0)v_2, z_2^{-1}) \\ & = z_1^{-1}\delta\left(\frac{z_2+z_0}{z_1}\right) Y(Y(e^{z_1L(1)}(-z_1^{-2})^{L(0)}v_1, -z_0/z_1z_2)\cdot \\ & \quad \cdot e^{z_2L(1)}(-z_2^{-2})^{L(0)}v_2, z_2^{-1}), \end{aligned} \tag{5.2.33}$$

or to

$$\begin{aligned} & Y(e^{z_2L(1)}(-z_2^{-2})^{L(0)}Y(v_1, z_0)v_2, z_2^{-1}) \\ & = Y(Y(e^{(z_2+z_0)L(1)}(-(z_2+z_0)^{-2})^{L(0)}v_1, -z_0/(z_2+z_0)z_2)\cdot \\ & \quad \cdot e^{z_2L(1)}(-z_2^{-2})^{L(0)}v_2, z_2^{-1}). \end{aligned} \tag{5.2.34}$$

If we can prove

$$\begin{aligned} & e^{z_2L(1)}(-z_2^{-2})^{L(0)}Y(v_1, z_0) \\ & = Y(e^{(z_2+z_0)L(1)}(-(z_2+z_0)^{-2})^{L(0)}v_1, -z_0/(z_2+z_0)z_2)\cdot \\ & \quad \cdot e^{z_2L(1)}(-z_2^{-2})^{L(0)} \end{aligned} \tag{5.2.35}$$

or equivalently, the conjugation formula

$$\begin{aligned} & e^{zL(1)}(-z^{-2})^{L(0)}Y(v, z_0)(-z^{-2})^{-L(0)}e^{-zL(1)} \\ & = Y(e^{(z+z_0)L(1)}(-(z+z_0)^{-2})^{L(0)}v, -z_0/(z+z_0)z) \end{aligned} \tag{5.2.36}$$

for any element v of a vertex operator algebra, where the operators act on the algebra itself, then we will be done. But for this, it is sufficient to prove the following lemma, which, for later use, we formulate more generally for modules:

LEMMA 5.2.3. *Let V be a vertex operator algebra, W a V-module. The following conjugation formulas hold on W :*

$$z^{L(0)}Y(v, z_0)z^{-L(0)} = Y(z^{L(0)}v, zz_0) \tag{5.2.37}$$

$$e^{zL(1)}Y(v, z_0)e^{-zL(1)} = Y(e^{z(1-zz_0)L(1)}(1-zz_0)^{-2L(0)}v, z_0/(1-zz_0)). \tag{5.2.38}$$

Proof. These formulas should be compared with the analogous result for $L(-1)$ – the outer equality in (4.2.11) – and with (2.6.7), which gives the $\mathfrak{sl}(2)$-conjugation action for quasiprimary fields (recall Section 4.4). Formula (5.2.37), which is similar to the case $n = 0$ of (2.6.7), follows from (4.2.2) and in fact is simply a reformulation of formula (4.2.4) for the weight of the operator v_n.

Indeed, equating the coefficients of the powers of z_0 on the two sides of (5.2.37) and assuming that v is homogeneous gives

$$z^{L(0)}v_n z^{-L(0)} = v_n z^{\mathrm{wt}\ v-n-1} \text{ for } n \in \mathbb{Z}. \tag{5.2.39}$$

Finally, we prove (5.2.38). (Note that if v is a lowest weight vector for $\mathfrak{sl}(2)$ (recall (2.6.9)), the factor $e^{z(1-zz_0)L(1)}$ drops out and (5.2.38) is the same as (2.6.7) for $n = 1$.) From (4.2.3) we find that

$$e^{zL(1)}Y(v, z_0)e^{-zL(1)} = Y(e^{zL(1)+2zz_0L(0)+zz_0^2L(-1)}v, z_0). \tag{5.2.40}$$

Using the second equality in (4.2.11), we see that it is sufficient to show that

$$\begin{aligned} &e^{zL(1)+2zz_0L(0)+zz_0^2L(-1)} \\ &\quad = e^{zz_0^2(1-zz_0)^{-1}L(-1)}e^{z(1-zz_0)L(1)}(1-zz_0)^{-2L(0)}, \end{aligned} \tag{5.2.41}$$

or that

$$\begin{aligned} &e^{zL(1)+2zz_0L(0)+zz_0^2L(-1)} \cdot \\ &\quad \cdot (1-zz_0)^{2L(0)}e^{-z(1-zz_0)L(1)}e^{-zz_0^2(1-zz_0)^{-1}L(-1)} = 1. \end{aligned} \tag{5.2.42}$$

For this, we compute the derivative of the left-hand side of (5.2.42) with respect to z and we use (5.2.12), (5.2.13), the first equality in (5.2.14), and (5.2.15). Setting

$$A = e^{zL(1)+2zz_0L(0)+zz_0^2L(-1)} \tag{5.2.43}$$

$$B = (1-zz_0)^{2L(0)} = e^{2\log(1-zz_0)L(0)} \tag{5.2.44}$$

$$C = e^{-z(1-zz_0)L(1)} \tag{5.2.45}$$

$$D = e^{-zz_0^2(1-zz_0)^{-1}L(-1)}, \tag{5.2.46}$$

we obtain

$$\begin{aligned} &A\left((L(1) + 2z_0L(0) + z_0^2L(-1))BCD - 2z_0(1-zz_0)^{-1}L(0)BCD \right. \\ &\quad \left. - (1-2zz_0)BL(1)CD - z_0^2(1-zz_0)^{-2}BCL(-1)D\ \right) \\ &= A\left((L(1) - 2zz_0^2(1-zz_0)^{-1}L(0) + z_0^2L(-1))BCD \right. \\ &\quad - (1-2zz_0)(1-zz_0)^{-2}L(1)BCD \\ &\quad - z_0^2(1-zz_0)^{-2}B(L(-1) - 2z(1-zz_0)L(0) \\ &\quad \left. + z^2(1-zz_0)^2L(1))CD\ \right) \\ &= A\left(z^2z_0^2(1-zz_0)^{-2}L(1) - 2zz_0^2(1-zz_0)^{-1}L(0) + z_0^2L(-1)\right)BCD \\ &\quad - Az_0^2\left(L(-1) - 2z(1-zz_0)^{-1}L(0) + z^2(1-zz_0)^{-2}L(1)\right)BCD \\ &= 0. \end{aligned}$$

Now we set $z = 0$ in the left-hand side of (5.2.42) to evaluate the constant, and this gives (5.2.42), proving Lemma 5.2.3 and hence Theorem 5.2.1.

5.3. Properties of contragredient modules

Let V be a vertex operator algebra, W a V-module.

PROPOSITION 5.3.1. *There are natural identifications between the double-contragredient module* W'' *and* W, *and between the double-adjoint operator* $Y''(\cdot, z)$ *and* $Y(\cdot, z)$.

Proof. There is a natural linear isomorphism identifying the double-graded-dual space W'' with W. It is sufficient to show that $Y''(v,z) = Y(v,z)$, i.e., that $Y(e^{z^{-1}L(1)}(-z^2)^{L(0)}e^{zL(1)}(-z^2)^{-L(0)}v, z) = Y(v,z)$ for $v \in V$. This will follow from the conjugation formula

$$(-z^2)^{L(0)}e^{zL(1)}(-z^2)^{-L(0)} = e^{-z^{-1}L(1)} \tag{5.3.1}$$

on V, and in turn from the formula

$$(-z^2)^{L(0)}zL(1)(-z^2)^{-L(0)} = -\frac{1}{z}L(1), \tag{5.3.2}$$

which realizes the transformation $z \mapsto -1/z$. But this follows from the fact that

$$z_0^{L(0)}L(1)z_0^{-L(0)} = \frac{1}{z_0}L(1), \tag{5.3.3}$$

which is immediately verified by applying to a homogeneous vector.

PROPOSITION 5.3.2. *The module* W *is irreducible if and only if* W' *is irreducible.*

Proof. By the last result, it is sufficient to show that if W is reducible, then W' is. But if W_1 is a (necessarily graded) proper nonzero submodule of W, then we see from the definition (5.2.4) that the annihilator of W_1 in W', necessarily proper and nonzero, is a submodule of W'.

Remark 5.3.3. Specifying a nondegenerate bilinear form $(\cdot,\cdot)$ on W such that

$$(W_{(m)}, W_{(n)}) = 0 \text{ for } m \neq n \tag{5.3.4}$$

is equivalent to specifying a grading-preserving linear isomorphism

$$\varphi : W \to W', \tag{5.3.5}$$

with the relationship given by

$$\langle \varphi w_1, w_2\rangle = (w_1, w_2) \text{ for } w_1, w_2 \in W. \tag{5.3.6}$$

It is clear that the map φ is a V-module isomorphism if and only if the form $(\cdot,\cdot)$ satisfies the following adjoint condition:

$$(Y(v,z)w_1, w_2) = (w_1, Y(e^{zL(1)}(-z^{-2})^{L(0)}v, z^{-1})w_2) \tag{5.3.7}$$

for $v \in V$, $w_1, w_2 \in W$. (Note also that this condition implies (5.3.4).) Thus W and W' are equivalent V-modules if and only if W admits a nondegenerate form with the indicated properties. More generally, given two V-modules W_1 and W_2, a V-module isomorphism

$$W_1 \to W_2' \tag{5.3.8}$$

amounts to a nondegenerate bilinear pairing

$$(\cdot,\cdot) : W_1 \times W_2 \to \mathbb{F} \tag{5.3.9}$$

satisfying the adjoint condition (5.3.7). Note that formula (5.3.1) implies that this adjoint condition holds on the opposite side as well:

$$(w_1, Y(v,z)w_2) = (Y(e^{zL(1)}(-z^{-2})^{L(0)}v, z^{-1})w_1, w_2) \tag{5.3.10}$$

for $w_1 \in W_1$, $w_2 \in W_2$.

Remark 5.3.4. Over $\mathbb{C}$, we may consider hermitian forms and hermitian pairings in place of bilinear ones, and the last remark and related comments below can easily be extended to include this situation.

Remark 5.3.5. The adjoint condition (5.3.7) (or (5.2.4)) can be rewritten a bit more simply if we use the following variant of the vertex operator $Y(\cdot, z)$ defining the module (W, Y) : For $v \in V$, set

$$X(v,z) = Y(z^{L(0)}v, z). \tag{5.3.11}$$

Note that if v is homogeneous, then

$$X(v,z) = z^{\mathrm{wt}\ v}Y(v,z). \tag{5.3.12}$$

The expansion coefficients of $X(v,z)$ are the operators defined in (4.2.5):

$$X(v,z) = \sum_{n\in\mathbb{Z}} x_v(n)z^{-n}, \tag{5.3.13}$$

even if v is not necessarily homogeneous, and $X(v,z)$ has the following simple transformation properties with respect to $L(0)$:

$$[L(0), X(v,z)] = z\frac{d}{dz}X(v,z) \tag{5.3.14}$$

$$z_0^{L(0)}X(v,z)z_0^{-L(0)} = X(v, z_0 z) \tag{5.3.15}$$

(cf. (4.2.6), (5.2.37)). With the help of the conjugation formula (5.3.3), we see that (5.3.7) becomes

$$(X(v,z)w_1, w_2) = (w_1, X(e^{L(1)}(-1)^{L(0)}v, z^{-1})w_2), \tag{5.3.16}$$

i.e.,

$$(X(v,z)w_1, w_2) = (-1)^{\mathrm{wt}\ v}(w_1, X(e^{L(1)}v, z^{-1})w_2) \tag{5.3.17}$$

in case v is homogeneous. If in addition v is a lowest weight vector for $\mathfrak{sl}(2)$ (i.e., $L(1)v = 0$), then the operator $e^{L(1)}$ drops out, and we have

$$(x_v(n)w_1, w_2) = (-1)^{\mathrm{wt}\ v}(w_1, x_v(-n)w_2) \quad \text{for } n \in \mathbb{Z}, \tag{5.3.18}$$

in agreement with (5.2.10), the case $v = \omega$.

The bilinear form $(\cdot,\cdot)$ on W discussed in Remark 5.3.3 is not assumed symmetric. However, it will typically turn out to be symmetric. For example, we have:

PROPOSITION 5.3.6. *Suppose that the contragredient V' of the adjoint module V is equivalent to V, and let $(\cdot,\cdot)$ be the nondegenerate bilinear form on V corresponding to a given isomorphism $V \to V'$. Then $(\cdot,\cdot)$ is symmetric.*

Proof. Let $v_1, v_2 \in V$. It is sufficient to show that

$$(Y(v_1,z)\mathbf{1}, v_2) = (v_2, Y(v_1,z)\mathbf{1}). \tag{5.3.19}$$

It will be little extra effort to compute directly that

$$(Y(v_1,z)v_0, v_2) = (v_2, Y(v_1,z)v_0) \tag{5.3.20}$$

for $v_i \in V$, and we shall want to refer to this computation later. In the following computation, we shall use skew-symmetry (2.3.19) for expressions based on each of the three pairs from the elements v_0, v_1 and v_2, and we shall use the adjoint formula (5.3.7) (or its reverse (5.3.10)) for expressions based on each of the three elements v_0, v_1 and v_2; the special case (5.2.10) of the adjoint formula is also used, together with the transformation formula (5.2.37) for the element v_1 and the general conjugation formula (5.3.1). We have

$$\begin{aligned}
(Y(v_1,z)v_0, v_2) &= (v_0, Y(e^{zL(1)}(-z^{-2})^{L(0)}v_1, z^{-1})v_2)\\
&= (v_0, e^{z^{-1}L(-1)}Y(v_2, -z^{-1})e^{zL(1)}(-z^{-2})^{L(0)}v_1)\\
&= (e^{z^{-1}L(1)}v_0, Y(v_2, -z^{-1})e^{zL(1)}(-z^{-2})^{L(0)}v_1)\\
&= (Y(e^{-z^{-1}L(1)}(-z^{-2})^{L(0)}v_2, -z)e^{z^{-1}L(1)}v_0, e^{zL(1)}(-z^{-2})^{L(0)}v_1)\\
&= (Y(e^{z^{-1}L(1)}v_0, z)e^{-z^{-1}L(1)}(-z^{2})^{L(0)}v_2, (-z^{-2})^{L(0)}v_1)\\
&= (e^{-z^{-1}L(1)}(-z^{2})^{L(0)}v_2, Y(e^{zL(1)}(-z^{-2})^{L(0)}\cdot\\
&\qquad \cdot e^{z^{-1}L(1)}v_0, z^{-1})(-z^{-2})^{L(0)}v_1)\\
&= ((-z^{2})^{L(0)}v_2, Y((-z^{-2})^{L(0)}v_1, -z^{-1})e^{zL(1)}(-z^{-2})^{L(0)}e^{z^{-1}L(1)}v_0)\\
&= (v_2, (-z^{2})^{L(0)}Y((-z^{-2})^{L(0)}v_1, -z^{-1})(-z^{-2})^{L(0)}v_0)\\
&= (v_2, Y(v_1,z)v_0),
\end{aligned} \tag{5.3.21}$$

as desired.

5.4. Intertwining operators

We add to the spaces listed in (2.1.1)-(2.1.5) the space

$$V\{z\} = \left\{ \sum_{n\mathbb{Q}} v_n z^n \middle| v_n \in V \right\} \tag{5.4.1}$$

of V-valued formal series involving the rational powers of z, for a vector space V.

DEFINITION 5.4.1. Let V be a vertex operator algebra and let (W_i, Y_i), (W_j, Y_j) and (W_k, Y_k) be three V-modules (not necessarily distinct, and possibly equal to V). An *intertwining operator of type* $\binom{i}{j\ k}$ (or *of type* $\binom{W_i}{W_j\ W_k}$) is a linear map $W_j \otimes W_k \to W_i\{z\}$, or equivalently,

$$\begin{aligned} W_j &\to (\mathrm{Hom}(W_k, W_i))\{z\} \\ w &\mapsto \mathcal{Y}(w, z) = \sum_{n \in \mathbb{Q}} w_n z^{-n-1} \quad (\text{where } w_n \in \mathrm{Hom}(W_k, W_i)) \end{aligned} \tag{5.4.2}$$

such that "all the defining properties of a module action that make sense hold" (cf. Definition 4.1.1). That is, for $v \in V$, $w_{(j)} \in W_j$ and $w_{(k)} \in W_k$,

$$w_{(j)n} w_{(k)} = 0 \text{ for } n \text{ sufficiently large;} \tag{5.4.3}$$

the following Jacobi identity holds for the operators $Y(v, \cdot)$, $\mathcal{Y}(w_{(j)}, \cdot)$ acting on the element $w_{(k)}$:

$$\begin{aligned} &z_0^{-1}\delta\left(\frac{z_1 - z_2}{z_0}\right) Y_i(v, z_1)\mathcal{Y}(w_{(j)}, z_2)w_{(k)} \\ &\quad - z_0^{-1}\delta\left(\frac{z_2 - z_1}{-z_0}\right) \mathcal{Y}(w_{(j)}, z_2)Y_k(v, z_1)w_{(k)} \\ &= z_2^{-1}\delta\left(\frac{z_1 - z_0}{z_2}\right) \mathcal{Y}(Y_j(v, z_0)w_{(j)}, z_2)w_{(k)} \end{aligned} \tag{5.4.4}$$

(note that the first term on the left-hand side is algebraically meaningful because of condition (5.4.3), and the other terms are meaningful by the usual properties of modules; also note that this Jacobi identity involves integral powers of z_0 and z_1 and rational powers of z_2);

$$\frac{d}{dz}\mathcal{Y}(w_{(j)}, z) = \mathcal{Y}(L(-1)w_{(j)}, z), \tag{5.4.5}$$

where $L(-1)$ is the operator acting on $W_{(j)}$. (Strictly speaking, we assume that the indices i, j and k are all distinct, since this allows us to use notations such as $w_{(i)}, w_{(j)}$ and $w_{(k)}$ without the requirement that say $w_{(i)} = w_{(j)}$ if $i \neq j$.)

We may denote the intertwining operator just defined by

$$\mathcal{Y}^i_{jk} \quad \text{or} \quad \mathcal{Y}^{W_i}_{W_j W_k}, \tag{5.4.6}$$

if necessary, to indicate its type.

Remark 5.4.2. Note that $Y(\cdot, z)$ acting on V is an example of an intertwining operator of type $\binom{V}{V\ V}$, and $Y(\cdot, z)$ acting on a V-module W is an example of an intertwining operator of type $\binom{W}{V\ W}$. These intertwining operators satisfy the normalization condition $Y(\mathbf{1}, z) = 1$. Also, the operator $Y(\cdot, z)$ defined by the skew-symmetry condition in (5.1.5) is an intertwining operator of type $\binom{W}{W\ V}$ (recall Proposition 5.1.2).

The intertwining operators of type $\binom{i}{j\ k}$ clearly form a vector space, which we denote by $\mathcal{V}^i_{jk}$ or $\mathcal{V}^{W_i}_{W_j W_k}$. We set

$$N^i_{jk} = N^{W_i}_{W_j W_k} = \dim \mathcal{V}^i_{jk} \ (\leq \infty). \tag{5.4.7}$$

These numbers are called the *fusion rules* associated with the algebra and modules. Then for example, assuming that V and the V-module W are nonzero, the corresponding fusion rules are positive:

$$N^V_{VV} \geq 1, \tag{5.4.8}$$

$$N^W_{VW} \geq 1, \tag{5.4.9}$$

$$N^W_{WV} \geq 1. \tag{5.4.10}$$

As in the case of module actions, we have the standard consequences:

$$[L(-1), \mathcal{Y}(w_{(j)}, z)] = \mathcal{Y}(L(-1)w_{(j)}, z) \tag{5.4.11}$$

$$[L(0), \mathcal{Y}(w_{(j)}, z)] = \mathcal{Y}(L(0)w_{(j)}, z) + z\mathcal{Y}(L(-1)w_{(j)}, z) \tag{5.4.12}$$

$$\begin{aligned} &[L(1), \mathcal{Y}(w_{(j)}, z)] \\ &= \mathcal{Y}(L(1)w_{(j)}, z) + 2z\mathcal{Y}(L(0)w_{(j)}, z) + z^2\mathcal{Y}(L(-1)w_{(j)}, z); \end{aligned} \tag{5.4.13}$$

note that each of the operators $L(-1)$, $L(0)$, $L(1)$ is acting here on three different spaces, and that on the left-hand sides, we are taking a liberty with bracket notation. If $w_{(j)} \in W_j$ is homogeneous, then

$$\text{wt } w_{(j)n} = \text{wt } w_{(j)} - n - 1 \ \text{ for } n \in \mathbb{Q}, \tag{5.4.14}$$

where the weight of a (homogeneous) operator from one graded space to another is defined in the obvious way. In particular, the operator $x_{w_{(j)}}(n)$ from W_k to W_i defined by

$$\mathcal{Y}(w_{(j)}, z) = \sum_{n \in \mathbb{Q}} x_{w_{(j)}}(n) z^{-n-\text{wt } w_{(j)}} \tag{5.4.15}$$

when $w_{(j)}$ is a homogeneous vector has weight $-n$:

$$\text{wt } x_{w_{(j)}}(n) = -n. \tag{5.4.16}$$

As usual, the notation $x_{w_{(j)}}(n)$ may be extended by linearity to arbitrary $w_{(j)} \in W_j$, and (5.4.16) is valid in general.

Taking Res_{z_0} of the Jacobi identity, we obtain the commutator formula (where we begin to drop the subscripts on the $Y's$ and where we continue to use bracket notation in a flexible way, as above)

$$[Y(v,z_1),\mathcal{Y}(w_{(j)},z_2)]=\mathrm{Res}_{z_0}z_2^{-1}\delta\left(\frac{z_1-z_0}{z_2}\right)\mathcal{Y}(Y(v,z_0)w_{(j)},z_2)$$
$$=\mathcal{Y}((Y(v,z_1-z_2)-Y(v,-z_2+z_1))w_{(j)},z_2) \tag{5.4.17}$$

as usual, and only the singular terms in $Y(v,z_0)w_{(j)}$ $(=Y_j(v,z_0)w_{(j)})$ enter into the commutator. For $v\in V$ and $w_{(j)}\in W_j$,

$$[v_0,\mathcal{Y}(w_{(j)},z)]=\mathcal{Y}(v_0w_{(j)},z) \tag{5.4.18}$$

$$[v_0,w_{(j)n}]=(v_0w_{(j)})_n \quad \text{for } n\in\mathbb{Q}, \tag{5.4.19}$$

$$[v_0,w_{(j)0}]=(v_0w_{(j)})_0. \tag{5.4.20}$$

Also,

$$e^{z_0L(-1)}\mathcal{Y}(w_{(j)},z)e^{-z_0L(-1)}=\mathcal{Y}(e^{z_0L(-1)}w_{(j)},z)=\mathcal{Y}(w_{(j)},z+z_0) \tag{5.4.21}$$

$$z^{L(0)}\mathcal{Y}(w_{(j)},z_0)z^{-L(0)}=\mathcal{Y}(z^{L(0)}w_{(j)},zz_0) \tag{5.4.22}$$

$$e^{zL(1)}\mathcal{Y}(w_{(j)},z_0)e^{-zL(1)}$$
$$=\mathcal{Y}(e^{z(1-zz_0)L(1)}(1-zz_0)^{-2L(0)}w_{(j)},z_0/(1-zz_0)) \tag{5.4.23}$$

(recall Lemma 5.2.3 and its proof).

Moreover, as in formulas (2.7.6) and (2.7.7), Res_{z_1} of the Jacobi identity gives:

$$\mathcal{Y}(Y(v,z_0)w_{(j)},z_2)=Y(v,z_0+z_2)\mathcal{Y}(w_{(j)},z_2)$$
$$+\mathcal{Y}(w_{(j)},z_2)(Y(v,z_2+z_0)-Y(v,z_0+z_2)). \tag{5.4.24}$$

As in Remark 5.3.5, we may define

$$\mathcal{X}(w_{(j)},z)=\mathcal{Y}(z^{L(0)}w_{(j)},z), \tag{5.4.25}$$

and we have

$$\mathcal{X}(w_{(j)},z)=\sum_{n\in\mathbb{Q}}x_{w_{(j)}}(n)z^{-n} \tag{5.4.26}$$

$$[L(0),\mathcal{X}(w_{(j)},z)]=z\frac{d}{dz}\mathcal{X}(w_{(j)},z) \tag{5.4.27}$$

$$z_0^{L(0)}\mathcal{Y}(w_{(j)}, z)z_0^{-L(0)} = \mathcal{Y}(w_{(j)}, z_0 z). \tag{5.4.28}$$

Remark 5.4.3. If W_i, W_j and/or W_k is a direct sum of submodules, then the space of intertwining operators is clearly the corresponding direct sum of the spaces of intertwining operators for the submodules, and the fusion rules correspondingly add.

Remark 5.4.4. If the modules W_i, W_j and W_k are irreducible, then from Remark 4.3.2 and (5.4.14) we see that we may refine (5.4.2):

$$\mathcal{Y}(w_{(j)}, z) \subset z^s \mathrm{Hom}(W_k, W_i)[[z, z^{-1}]] \text{ for } w_{(j)} \in W_j, \tag{5.4.29}$$

with s a rational number given by:

$$s = \mathrm{wt}\ w_{(i)} - \mathrm{wt}\ w_{(j)} - \mathrm{wt}\ w_{(k)}, \tag{5.4.30}$$

where $w_{(i)}$ and $w_{(k)}$ are arbitrary nonzero homogeneous elements of W_i and W_k, respectively (and where $w_{(j)}$ is taken to be nonzero and homogeneous). In particular, $\mathcal{Y}(w_{(j)}, z)$ is of the form z^s times a formal Laurent series involving only integral powers of z.

Remark 5.4.5. The discussion of primary and quasiprimary fields of Section 2.6 holds for intertwining operators, and in particular, if $w_{(j)} \in W_j$ is a lowest weight vector for the Virasoro algebra (respectively, for $\mathfrak{sl}(2)$), then $\mathcal{Y}(w_{(j)}, z)$ is a primary (respectively, quasiprimary) field (cf. Section 4.4).

Remark 5.4.6. An intertwining operator with a given restriction to some subset S of W_j is determined on the V-submodule of W_j generated by S, in view of (5.4.24). For example, if W_j is irreducible, then since it is generated by a (nonzero) weight vector $w_{(j)}$ with the lowest possible weight, the intertwining operator $\mathcal{Y}$ is determined by its associated primary field $\mathcal{Y}(w_{(j)}, z)$. Analogously, by the commutator formula (5.4.17), the intertwining operator $\mathcal{Y}$ is determined by its action on a lowest weight vector of W_k (or on any nonzero element of W_k) in case W_k is irreducible.

The fusion rules have various symmetry properties. Here we establish one such property, and in the next section, another one. In both cases we make certain integrality assumptions for convenience, so as to avoid the need to interpret rational powers of -1.

Let $\mathcal{Y}$ be an intertwining operator of type $\binom{i}{j\ k}$, in the notation of Definition 5.4.1. We suppose that $\mathcal{Y}(\cdot, z)$ involves only integral powers of z, as in the cases of the operators Y recalled in Remark 5.4.2: For $w_{(j)} \in W_j$, $w_{(k)} \in W_k$,

$$\mathcal{Y}(w_{(j)}, z)w_{(k)} \in W_i[[z, z^{-1}]]. \tag{5.4.31}$$

We use the skew-symmetry condition (2.3.19) to construct another operator

$$\mathcal{Y}^* : W_k \otimes W_j \to W_i[[z, z^{-1}]] \tag{5.4.32}$$

by the formula

$$\mathcal{Y}^*(w_{(k)}, z)w_{(j)} = e^{zL(-1)}\mathcal{Y}(w_{(j)}, -z)w_{(k)}. \tag{5.4.33}$$

We verify that $\mathcal{Y}^*$ is an intertwining operator of type $\binom{i}{k\ j}$: First, formula (5.4.3) is clear. Next, the second half of the proof of the $\mathcal{S}_3$-symmetry of the Jacobi identity, Proposition 2.7.1, gives us the Jacobi identity (5.4.4). Finally, as in the setting of (5.1.6), we note that (5.4.21) implies the corresponding formula for $\mathcal{Y}^*$:

$$\begin{aligned} e^{z_0L(-1)}\mathcal{Y}^*(w_{(k)}, z)e^{-z_0L(-1)} &= \mathcal{Y}^*(e^{z_0L(-1)}w_{(k)}, z) \\ &= \mathcal{Y}^*(w_{(k)}, z + z_0), \end{aligned} \tag{5.4.34}$$

which in turn implies (5.4.5). Thus we have:

PROPOSITION 5.4.7. *The operator $\mathcal{Y}^*$ is an intertwining operator of type $\binom{i}{k\ j}$. Moreover,*

$$\mathcal{Y}^{**} = \mathcal{Y}. \tag{5.4.35}$$

In particular, suppose the three modules W_i, W_j, W_k satisfy the condition that

$$\mathrm{wt}\ w_{(i)} - \mathrm{wt}\ w_{(j)} - \mathrm{wt}\ w_{(k)} \in \mathbb{Z} \tag{5.4.36}$$

for all homogeneous elements $w_{(i)} \in W_i$, $w_{(j)} \in W_j$, $w_{(k)} \in W_k$ (cf. (5.4.30)), so that

$$\mathcal{Y}(w_{(j)}, z)w_{(k)} \in W_i[[z, z^{-1}]] \tag{5.4.37}$$

for every intertwining operator $\mathcal{Y}$ of type $\binom{i}{j\ k}$. Then the correspondence $\mathcal{Y} \mapsto \mathcal{Y}^*$ defines a linear isomorphism from $\mathcal{V}^i_{jk}$ to $\mathcal{V}^i_{kj}$, and we have

$$N^i_{jk} = N^i_{kj}. \tag{5.4.38}$$

5.5. Adjoint intertwining operators

Let V be a vertex operator algebra, and let (W_i, Y_i), (W_j, Y_j) and (W_k, Y_k) be three V-modules, not necessarily distinct, and possibly equal to V, as in Section 5.4. We are interested in defining the adjoint operator of an intertwining operator. However, the formula for the adjoint involves the expression $(-1)^{L(0)}$, and in order to make sense of this expression for arbitrary $\mathbb{Q}$-graded modules we would have to choose suitable roots of unity. Since such considerations are not important for our main goals, we shall make appropriate integrality assumptions, for convenience, as in Proposition 5.4.7.

Suppose, then, that the grading of W_j is integral:

$$W_j = \coprod_{n\in\mathbb{Z}} (W_j)_{(n)} \tag{5.5.1}$$

Given an intertwining operator

$$\begin{aligned} &W_j \otimes W_k \to W_i\{z\} \\ &w_{(j)} \otimes w_{(k)} \mapsto \mathcal{Y}(w_{(j)}, z)w_{(k)} = \mathcal{Y}^i_{jk}(w_{(j)}, z)w_{(k)}, \end{aligned} \tag{5.5.2}$$

the *adjoint operator of* $\mathcal{Y}$ is the linear map

$$\begin{aligned} W_j \otimes W_i' &\to W_k'\{z\} \\ w_{(j)} \otimes w'_{(i)} &\mapsto \mathcal{Y}'(w_{(j)}, z)w'_{(i)} = (\mathcal{Y}^i_{jk})'(w_{(j)}, z)w'_{(i)} \end{aligned} \tag{5.5.3}$$

defined by

$$\begin{aligned} &\langle \mathcal{Y}'(w_{(j)}, z)w'_{(i)}, w_{(k)}\rangle_k \\ &= \langle w'_{(i)}, \mathcal{Y}(e^{zL(1)}(-z^{-2})^{L(0)}w_{(j)}, z^{-1})w_{(k)}\rangle_i, \end{aligned} \tag{5.5.4}$$

where $\langle \cdot, \cdot \rangle_k$ and $\langle \cdot, \cdot \rangle_i$ are the pairings for W_k, W_k' and W_i, W_i', respectively; then $\mathcal{Y}'$ is well defined, just as in (5.2.3)-(5.2.4) (this time, using (5.4.14)).

THEOREM 5.5.1. *The adjoint operator of an intertwining operator (of type* $\binom{W_i}{W_j W_k}$ *or* $\binom{i}{j\ k}$*) is an intertwining operator (of type* $\binom{W_k'}{W_j W_i'}$ *or, for brevity,* $\binom{k'}{j\ i'}$*.)*

Proof. The proof is the same as the relevant portion of the proof of Theorem 5.2.1, except of course that when we use (5.5.4) instead of (5.2.4) we keep in mind that the pairings on the two sides of (5.5.4) are for different spaces; also, Lemma 5.2.3 is used here in the generality of modules.

Recall from Proposition 5.3.1 that a double-contragredient module is naturally identified with the original module. The proof of that result, with formulas (5.3.1)-(5.3.3) understood as identities on W_j, gives:

PROPOSITION 5.5.2. *The double-adjoint operator* $\mathcal{Y}''$ *equals the intertwining operator* $\mathcal{Y}$*. In particular, the correspondence* $\mathcal{Y} \mapsto \mathcal{Y}'$ *defines a linear isomorphism from* $\mathcal{V}^i_{jk}$ *to* $\mathcal{V}^{k'}_{ji'}$*, and we have*

$$N^i_{jk} = N^{k'}_{ji'} \tag{5.5.5}$$

(using obvious notation).

In view of these propositions, we might wish to use the notation

$$\mathcal{Y}' = (\mathcal{Y}^i_{jk})' = \mathcal{Y}^{k'}_{ji'}. \tag{5.5.6}$$

Note that if we set

$$N_{ijk} = N^{i'}_{jk}, \tag{5.5.7}$$

then under the integrality assumptions of Propositions 5.4.7 and 5.5.2, the N_{ijk}'s are S_3-symmetric:

$$N_{ijk} = N_{\sigma(i)\sigma(j)\sigma(k)} \tag{5.5.8}$$

for every permutation σ of $\{i, j, k\}$.

Combining Propositions 5.5.1 and 5.5.2 and Remark 5.3.3, we immediately obtain the following:

COROLLARY 5.5.3. *Assume that we have nondegenerate bilinear forms* $(\cdot,\cdot)_i$ *and* $(\cdot,\cdot)_k$ *on* W_i *and* W_k*, respectively, such that the two adjoint properties*

$$(Y_i(v,z)\tilde{w}_{(i)},w_{(i)})_i=(\tilde{w}_{(i)},Y_i(e^{zL(1)}(-z^{-2})^{L(0)}v,z^{-1})w_{(i)})_i, \tag{5.5.9}$$

$$(Y_k(v,z)\tilde{w}_{(k)},w_{(k)})_k=(\tilde{w}_{(k)},Y_k(e^{zL(1)}(-z^{-2})^{L(0)}v,z^{-1})w_{(k)})_k \tag{5.5.10}$$

hold for $v\in V$, $w_{(i)},\tilde{w}_{(i)}\in W_i$, $w_{(k)},\tilde{w}_{(k)}\in W_k$. *Also assume that we have an intertwining operator* $\mathcal{Y}^i_{jk}$ *of type* $\binom{i}{j\ k}$*, as in (5.5.2). Define an adjoint operator*

$$\begin{aligned}W_j\otimes W_i&\to W_k\{z\}\\ w_{(j)}\otimes w_{(i)}&\mapsto \mathcal{Y}^k_{ji}(w_{(j)},z)w_{(i)}\end{aligned} \tag{5.5.11}$$

by the formula

$$\begin{aligned}&(\mathcal{Y}^k_{ji}(w_{(j)},z)w_{(i)},w_{(k)})_k\\ &=(w_{(i)},\mathcal{Y}(e^{zL(1)}(-z^{-2})^{L(0)}w_{(j)},z^{-1})w_{(k)})_i.\end{aligned} \tag{5.5.12}$$

Equivalently, set

$$\mathcal{Y}^k_{ji}(w_{(j)},z)w_{(i)}=\varphi_k^{-1}(\mathcal{Y}'(w_{(j)},z)\varphi_i(w_{(i)})), \tag{5.5.13}$$

where $\varphi_i: W_i\to W_i'$ *and* $\varphi_k: W_k\to W_k'$ *are the module isomorphisms associated with* $(\cdot,\cdot)_i$ *and* $(\cdot,\cdot)_k$ *(see Remark 5.3.3). Then* $\mathcal{Y}^k_{ji}$ *is an intertwining operator of type* $\binom{k}{j\ i}$*, and the correspondence* $\mathcal{Y}^i_{jk}\mapsto\mathcal{Y}^k_{ji}$ *defines a linear isomorphism from* $\mathcal{V}^i_{jk}$ *to* $\mathcal{V}^k_{ji}$*. In particular,*

$$N^i_{jk}=N^k_{ji} \tag{5.5.14}$$

Remark 5.5.4. Theorem 5.5.1, Proposition 5.5.2 and Corollary 5.5.3 have the following obvious generalization: Assume that we have a vertex operator algebra V and five V-modules W_i, W_j, W_k, $\bar{W}_i$ and $\bar{W}_k$, with the grading of W_j integral. Also assume that we have an intertwining operator $\mathcal{Y}$ as in (5.5.2) and two nondegenerate bilinear forms

$$(\cdot,\cdot)_i:\bar{W}_i\times W_i\to\mathbb{F},\quad (\cdot,\cdot)_k:\bar{W}_k\times W_k\to\mathbb{F}, \tag{5.5.15}$$

each of which satisfies the usual adjoint condition (cf. (5.3.8)-(5.3.9)). We can define a generalized adjoint operator

$$\begin{aligned}W_j\otimes\bar{W}_i&\to\bar{W}_k\{z\}\\ w_{(j)}\otimes\bar{w}_{(i)}&\mapsto\mathcal{Y}^{\bar{k}}_{j\bar{i}}(w_{(j)},z)\bar{w}_{(i)}\end{aligned} \tag{5.5.16}$$

by

$$\begin{aligned}&(\mathcal{Y}^{\bar{k}}_{j\bar{i}}(w_{(j)},z)\bar{w}_{(i)},w_{(k)})_k\\ &=(\bar{w}_{(i)},\mathcal{Y}(e^{zL(1)}(-z^{-2})^{L(0)}w_{(j)},z^{-1})w_{(k)})_i,\end{aligned} \tag{5.5.17}$$

where $w_{(j)}\in W_j$, $w_{(k)}\in W_k$, $\bar{w}_{(i)}\in\bar{W}_i$. Then this operator $\mathcal{Y}^{\bar{k}}_{j\bar{i}}$ is an intertwining operator, and we have the corresponding isomorphism of spaces of intertwining operators and equality of fusion rules, again by Propositions 5.5.1 and 5.5.2 and Remark 5.3.3. Note that we might have $\bar{W}_i=W_i$, $\bar{W}_k=W_k$, as in Proposition 5.5.3, or we might have, say, $\bar{W}_i=W_k$, $\bar{W}_k=W_i$.

5.6. Duality for two module elements and one algebra element

We have discussed duality for one module element and two algebra elements in Section 5.1. Here, using the results of the previous several sections, we formulate and derive duality for two module elements and one algebra element.

Let V be a vertex operator algebra and let W be a V-module, which we assume to be $\mathbb{Z}$-graded:

$$W = \coprod_{n\in\mathbb{Z}} W_{(n)} \tag{5.6.1}$$

(cf. (5.5.1)). We already have three fixed intertwining operators – the algebra action, the module action and the operator given by skew-symmetry in (5.1.5) – which we shall denote

$$Y_{VV}^{V},\ Y_{VW}^{W},\ Y_{WV}^{W}, \tag{5.6.2}$$

respectively (recall Remark 5.4.2).

Suppose now that there are V-module equivalences

$$V \simeq V',\quad W \simeq W', \tag{5.6.3}$$

so that we have nondegenerate bilinear forms $(\cdot,\cdot)_V$ and $(\cdot,\cdot)_W$ on V and W, respectively, which satisfy the adjoint condition (5.3.7) with respect to Y_{VV}^{V} and Y_{VW}^{W}, respectively. Then we obtain an intertwining operator

$$\begin{aligned} W\otimes W &\to V[[z,z^{-1}]] \\ w_1\otimes w_2 &\mapsto Y_{WW}^{V}(w_1,z)w_2 \end{aligned} \tag{5.6.4}$$

by taking $W_i = W_j = W$, $W_k = V$ and $\mathcal{Y}_{jk}^{i} = Y_{WV}^{W}$ in Proposition 5.5.3. Explicitly, Y_{WW}^{V} is given by:

$$(Y_{WW}^{V}(w_1,z)w_2,v)_V = (w_2, Y_{WV}^{W}(e^{zL(1)}(-z^{-2})^{L(0)}w_1,z^{-1})v)_W \tag{5.6.5}$$

for $w_1,w_2\in W$, $v\in V$.

Recall from Proposition 5.3.6 that the form $(\cdot,\cdot)_V$ is symmetric. The proof used the adjoint formula (5.3.7) and the skew-symmetry formula (2.3.19). Observe now that in addition to (5.6.5), the adjoint formula holds for $(\cdot,\cdot)_W$ with respect to Y_{VW}^{W}, and that the skew-symmetry formula (5.1.5) holds, relating Y_{VW}^{W} and Y_{WV}^{W}. The proof of Proposition 5.3.6, with the elements v_0, v_1, v_2 in that proof taken to be in W, V, W, respectively, gives:

PROPOSITION 5.6.1. *The form $(\cdot,\cdot)_W$ is symmetric if and only if*

$$Y_{WW}^{V}(w_1,z)w_2 = e^{zL(-1)}Y_{WW}^{V}(w_2,-z)w_1 \tag{5.6.6}$$

for w_1, $w_2\in W$.

Let us assume the conditions of Proposition 5.6.1. Recall from Section 2.7 the notion of the Jacobi identity for an ordered triple of elements. (We may of course use the common notation Y for all the four intertwining operators Y_{VV}^{V}, Y_{VW}^{W},

Y^W_{WV}, Y^V_{WW}.) As above, let $v \in V$ and $w_1, w_2 \in W$. The fact that Y^V_{WW} is an intertwining operator tells us that the Jacobi identity holds for the ordered triple (v, w_1, w_2) and that the $L(-1)$-derivative property (5.4.5) holds. In particular, both equalities in (2.3.17) hold for Y^V_{WW}, as they do for each of our intertwining operators. Also, skew-symmetry (2.3.19) holds when both elements are in V, when both elements are in W, and when one element is in each. Thus the proof of Proposition 2.7.1 shows that the Jacobi identity holds for each permutation of the triple (v, w_1, w_2), that is, it holds for two elements of W and one element of V, where the element of V may be in any of the three positions. Then just as in Sections 3.2, 3.3 and 5.1, we conclude that rationality, commutativity and associativity hold for these same ordered triples. The proof of Proposition 3.5.1 holds as well (the induction hypothesis in that proof involving at most two module elements), and we conclude:

THEOREM 5.6.2. *Let V be a vertex operator algebra and let W be a V-module. Assume that W is $\mathbb{Z}$-graded, that there are V-module equivalences $V \simeq V'$, $W \simeq W'$, and that the resulting nondegenerate bilinear form on W is symmetric. Define operators $Y = Y^W_{WV}$ and $Y = Y^V_{WW}$ by the skew-symmetry condition (5.1.5) and the adjoint condition (5.6.5), respectively. Let u_1, u_2, u_3 be elements of either V or W, with at most two in W. Then the Jacobi identity as well as rationality, commutativity, associativity and the analytic assertions of Corollary 3.3.3 hold for the triple (u_1, u_2, u_3). Moreover, for elements $u_1, ..., u_n$ of either V or W, with at most two in W, rationality and commutativity hold for the n-tuple $(u_1, ..., u_n)$.*

Remark 5.6.3. The methods discussed in this work do not prove duality for more than two module elements, and indeed, such duality will not hold without additional hypotheses. Duality for three module elements would amount to the assertion that $V \oplus W$ carries the structure of a vertex operator algebra. This has been established for the moonshine module [FLM], our motivating example.

REFERENCES

[BPZ] A. A. Belavin, A. M. Polyakov and A. B. Zamolodchikov, *Infinite conformal symmetries in two-dimensional quantum field theory*, Nucl. Phys. **B241** (1984), 333–380.

[B] R. E. Borcherds, *Vertex algebras, Kac-Moody algebras, and the Monster*, Proc. Natl. Acad. Sci. USA **83** (1986), 3068-3071.

[DGM] L. Dolan, P. Goddard and P. Montague, *Conformal field theory of twisted vertex operators*, Nucl. Phys. **B338** (1990), 529-601.

[FLM] I. Frenkel, J. Lepowsky and A. Meurman, *Vertex Operator Algebras and the Monster*, Pure and Appl. Math., vol. 134, Academic Press, Boston, 1988.

[G] P. Goddard, *Meromorphic conformal field theory*, in: Infinite Dimensional Lie Algebras and Lie Groups: Proceedings of the CIRM-Luminy Conference, 1988, World Scientific, Singapore, 1989, p. 556.

[J] N. Jacobson, *Structure of Rings*, American Math. Soc. Colloquium Publ. vol. 37, American Math. Soc., Providence, RI, 1964.

[MS] G. Moore and N. Seiberg, *Classical and quantum conformal field theory*, Comm. Math. Phys. **123** (1989), 177-254.

[TK] A. Tsuchiya and Y. Kanie, *Vertex operators in conformal field theory on* $\mathbb{P}^1$ *and monodromy representations of braid group*, in: Conformal Field Theory and Solvable Lattice Models, Advanced Studies in Pure Math., vol. 16, Kinokuniya Company Ltd., Tokyo, 1988, pp. 297-372.

[V] E. Verlinde, *Fusion rules and modular transformations in 2D conformal field theory*, Nucl. Phys. **B300** (1988), 360-376.

Department of Mathematics, Yale University, New Haven, CT 06520

Department of Mathematics, Rutgers University, New Brunswick, NJ 08903

Current address: Department of Mathematics, University of Pennsylvania, Philadelphia, PA 19104

E-mail address: yzhuang@math.upenn.edu

Department of Mathematics, Rutgers University, New Brunswick, NJ 08903

E-mail address: lepowsky@math.rutgers.edu

Editorial Information

To be published in the *Memoirs*, a paper must be correct, new, nontrivial, and significant. Further, it must be well written and of interest to a substantial number of mathematicians. Piecemeal results, such as an inconclusive step toward an unproved major theorem or a minor variation on a known result, are in general not acceptable for publication. *Transactions* Editors shall solicit and encourage publication of worthy papers. Papers appearing in *Memoirs* are generally longer than those appearing in *Transactions* with which it shares an editorial committee.

As of May 6, 1993, the backlog for this journal was approximately 8 volumes. This estimate is the result of dividing the number of manuscripts for this journal in the Providence office that have not yet gone to the printer on the above date by the average number of monographs per volume over the previous twelve months, reduced by the number of issues published in four months (the time necessary for preparing an issue for the printer). (There are 6 volumes per year, each containing at least 4 numbers.)

A Copyright Transfer Agreement is required before a paper will be published in this journal. By submitting a paper to this journal, authors certify that the manuscript has not been submitted to nor is it under consideration for publication by another journal, conference proceedings, or similar publication.

Information for Authors and Editors

Memoirs are printed by photo-offset from camera copy fully prepared by the author. This means that the finished book will look exactly like the copy submitted.

The paper must contain a *descriptive title* and an *abstract* that summarizes the article in language suitable for workers in the general field (algebra, analysis, etc.). The *descriptive title* should be short, but informative; useless or vague phrases such as "some remarks about" or "concerning" should be avoided. The *abstract* should be at least one complete sentence, and at most 300 words. Included with the footnotes to the paper, there should be the 1991 *Mathematics Subject Classification* representing the primary and secondary subjects of the article. This may be followed by a list of *key words and phrases* describing the subject matter of the article and taken from it. A list of the numbers may be found in the annual index of *Mathematical Reviews*, published with the December issue starting in 1990, as well as from the electronic service e-MATH [**telnet e-MATH.ams.org** (or **telnet 130.44.1.100**). Login and password are **e-math**]. For journal abbreviations used in bibliographies, see the list of serials in the latest *Mathematical Reviews* annual index. When the manuscript is submitted, authors should supply the editor with electronic addresses if available. These will be printed after the postal address at the end of each article.

Electronically prepared manuscripts. The AMS encourages submission of electronically prepared manuscripts in $\mathcal{A}_{\mathcal{M}}\mathcal{S}$-TEX or $\mathcal{A}_{\mathcal{M}}\mathcal{S}$-LATEX because properly prepared electronic manuscripts save the author proofreading time and move more quickly through the production process. To this end, the Society has prepared "preprint" style files, specifically the amsppt style of $\mathcal{A}_{\mathcal{M}}\mathcal{S}$-TEX and the amsart style of $\mathcal{A}_{\mathcal{M}}\mathcal{S}$-LATEX, which will simplify the work of authors and of the production staff. Those authors who make use of these style files from the beginning of the writing process will further reduce their own effort. Electronically submitted manuscripts prepared in plain TEX or LATEX do not mesh properly with the AMS production systems and cannot, therefore, realize the same kind of expedited processing. Users of plain TEX should have little difficulty learning AMS-TEX, and LATEX users will find that AMS-LATEX is the same as LATEX with additional commands to simplify the typesetting of mathematics.

Guidelines for Preparing Electronic Manuscripts provides additional assistance and is available for use with either AMS-TEX or AMS-LATEX. Authors with FTP access may obtain *Guidelines* from the Society's Internet node e-MATH@math.ams.org (130.44.1.100). For those without FTP access *Guidelines* can be obtained free of charge from the e-mail address guide-elec@math.ams.org (Internet) or from the Publications Department, American Mathematical Society, P.O. Box 6248, Providence, RI 02940-6248. When requesting *Guidelines*, please specify which version you want.

At the time of submission, authors should indicate if the paper has been prepared using AMS-TEX or AMS-LATEX. The *Manual for Authors of Mathematical Papers* should be consulted for symbols and style conventions. The *Manual* may be obtained free of charge from the e-mail address cust-serv@math.ams.org or from the Customer Services Department, American Mathematical Society, P.O. Box 6248, Providence, RI 02940-6248. The Providence office should be supplied with a manuscript that corresponds to the electronic file being submitted.

Electronic manuscripts should be sent to the Providence office immediately after the paper has been accepted for publication. They can be sent via e-mail to pub-submit@math.ams.org (Internet) or on diskettes to the Publications Department address listed above. When submitting electronic manuscripts please be sure to include a message indicating in which publication the paper has been accepted. No corrections will be accepted electronically. Authors must mark their changes on their proof copies and return them to the Providence office. Authors and editors are encouraged to make the necessary submissions of electronically prepared manuscripts and proof copies in a timely fashion.

Two copies of the paper should be sent directly to the appropriate Editor and the author should keep one copy. The *Guide for Authors of Memoirs* gives detailed information on preparing papers for *Memoirs* and may be obtained free of charge from AMS, Editorial Department, P. O. Box 6248, Providence, RI 02940-6248. For papers not prepared electronically, model paper may also be obtained free of charge from the Editorial Department.

Any inquiries concerning a paper that has been accepted for publication should be sent directly to the Editorial Department, American Mathematical Society, P. O. Box 6248, Providence, RI 02940-6248.

Editors

Recent Titles in This Series

(Continued from the front of this publication)

462 **Toshiyuki Kobayashi,** Singular unitary representations and discrete series for indefinite Stiefel manifolds $U(p,q;\mathbb{F})/U(p-m,q;\mathbb{F})$, 1992

461 **Andrew Kustin and Bernd Ulrich,** A family of complexes associated to an almost alternating map, with application to residual intersections, 1992

460 **Victor Reiner,** Quotients of coxeter complexes and P-partitions, 1992

459 **Jonathan Arazy and Yaakov Friedman,** Contractive projections in C_p, 1992

458 **Charles A. Akemann and Joel Anderson,** Lyapunov theorems for operator algebras, 1991

457 **Norihiko Minami,** Multiplicative homology operations and transfer, 1991

456 **Michał Misiurewicz and Zbigniew Nitecki,** Combinatorial patterns for maps of the interval, 1991

455 **Mark G. Davidson, Thomas J. Enright and Ronald J. Stanke,** Differential operators and highest weight representations, 1991

454 **Donald A. Dawson and Edwin A. Perkins,** Historical processes, 1991

453 **Alfred S. Cavaretta, Wolfgang Dahmen, and Charles A. Micchelli,** Stationary subdivision, 1991

452 **Brian S. Thomson,** Derivates of interval functions, 1991

451 **Rolf Schön,** Effective algebraic topology, 1991

450 **Ernst Dieterich,** Solution of a non-domestic tame classification problem from integral representation theory of finite groups ($\Lambda = RC_3, v(3) = 4$), 1991

449 **Michael Slack,** A classification theorem for homotopy commutative H-spaces with finitely generated mod 2 cohomology rings, 1991

448 **Norman Levenberg and Hiroshi Yamaguchi,** The metric induced by the Robin function, 1991

447 **Joseph Zaks,** No nine neighborly tetrahedra exist, 1991

446 **Gary R. Lawlor,** A sufficient criterion for a cone to be area-minimizing, 1991

445 **S. Argyros, M. Lambrou, and W. E. Longstaff,** Atomic Boolean subspace lattices and applications to the theory of bases, 1991

444 **Haruo Tsukada,** String path integral realization of vertex operator algebras, 1991

443 **D. J. Benson and F. R. Cohen,** Mapping class groups of low genus and their cohomology, 1991

442 **Rodolfo H. Torres,** Boundedness results for operators with singular kernels on distribution spaces, 1991

441 **Gary M. Seitz,** Maximal subgroups of exceptional algebraic groups, 1991

440 **Bjorn Jawerth and Mario Milman,** Extrapolation theory with applications, 1991

439 **Brian Parshall and Jian-pan Wang,** Quantum linear groups, 1991

438 **Angelo Felice Lopez,** Noether-Lefschetz theory and the Picard group of projective surfaces, 1991

437 **Dennis A. Hejhal,** Regular b-groups, degenerating Riemann surfaces, and spectral theory, 1990

436 **J. E. Marsden, R. Montgomery, and T. Ratiu,** Reduction, symmetry, and phase mechanics, 1990

435 **Aloys Krieg,** Hecke algebras, 1990

434 **François Treves,** Homotopy formulas in the tangential Cauchy-Riemann complex, 1990

433 **Boris Youssin,** Newton polyhedra without coordinates Newton polyhedra of ideals, 1990

432 **M. W. Liebeck, C. E. Praeger, and J. Saxl,** The maximal factorizations of the finite simple groups and their automorphism groups, 1990

(See the AMS catalog for earlier titles)